供中药、药学类专业用

双语版

仪器分析实验

Instrument Analysis Experiment

Bilingual Edition

主审　郑荣庆　谢晓梅
主编　汪电雷　程旺兴

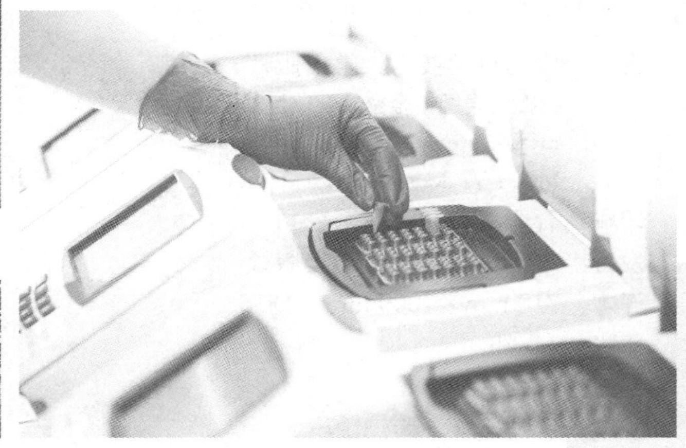

中国科学技术大学出版社

内 容 简 介

为了加快教育改革的步伐、培养具有国际竞争力的高水平药学类人才、加强学生在专业上英语的实际应用能力、满足国际生的交流和学习，结合高等医药院校的特色和仪器分析学科的特点，组织编写了本教材。

本教材是在长期教学、科研和教学实践基础上编写而成的，既能满足学生个性发展需要，又注重培养学生分析问题、解决问题的能力。全书共4章，包括仪器分析实验基本知识，验证性仪器分析实验13个，综合性仪器分析实验6个，设计性仪器分析实验3个。

本教材可供全国中医药院校中药、药学类专业及相关专业学生使用，也可为药学类相关研究工作者提供参考。

图书在版编目(CIP)数据

仪器分析实验：双语版：汉文、英文/汪电雷，程旺兴主编. —合肥：中国科学技术大学出版社，2023.6

ISBN 978-7-312-05686-4

Ⅰ. 仪… Ⅱ. ① 汪…② 程… Ⅲ. 仪器分析—实验—高等学校—教材—汉、英 Ⅳ. O657-33

中国国家版本馆CIP数据核字(2023)第087588号

仪器分析实验(双语版)

YIQI FENXI SHIYAN (SHUANGYU BAN)

出版 中国科学技术大学出版社
安徽省合肥市金寨路96号，230026
http://press.ustc.edu.cn
https://zgkxjsdxcbs.tmall.com

印刷 合肥皖科印务有限公司

发行 中国科学技术大学出版社

开本 710 mm×1000 mm 1/16

印张 11.75

字数 241千

版次 2023年6月第1版

印次 2023年6月第1次印刷

定价 36.00元

本书编委会

主　审　郑荣庆　谢晓梅

主　编　汪电雷　程旺兴

副主编　马　涛　陈云艳　朱丽丽

编　委（按姓名笔画排序）

马　涛（蚌埠医学院）
王　雷（安徽中医药大学）
王晓群（安徽中医药大学）
朱丽丽（安徽医科大学）
刘寒蒙（安徽中医药大学）
阮班峰（合肥学院）
芦男男（安徽中医药大学）
杨　晔（安徽中医药大学）
杨　敏（安徽中医药大学）
束成林（安徽中医药大学）
汪电雷（安徽中医药大学）
张　玲（安徽中医药大学）
张云静（安徽中医药大学）
张自品（安徽中医药大学）
陈　阳（蚌埠医学院）
陈乃东（皖西学院）
陈云艳（皖南医学院）
栗进才（亳州学院）
倪　佳（安徽中医药大学）
韩智莉（安徽中医药大学）
程旺兴（安徽中医药大学）
蔡百祥（安徽中医药大学）
薛　璇（安徽中医药大学）

前　言

本书紧密围绕国家教育部有关高校实验教学改革的要求和地方高水平大学建设目标要求，旨在提高药学类专业学生的实践动手能力，更好地推进教学内容、教学方法、教学手段改革的进程；同时加强学生的英语在专业上的实际应用能力，培养具有国际竞争力的高水平药学类人才，并促进国际生的学习和交流。结合高等医药院校的特色和仪器分析学科的特点，编委会根据药学类专业“仪器分析实验”课程教学大纲要求，结合当前实验教学仪器设备和条件，组织编写了本教材，供药学、中药学、药物制剂学和制药工程学等专业本科“仪器分析实验”课程教学使用。

本书在内容上力求充分吸取先进的实验技术和手段，根据仪器分析方法内在的规律和联系，以介绍分析仪器的使用为主，阐释其构造和原理为辅，注重基本实验操作能力的锻炼，先易后难，循序渐进。在培养良好的仪器规范使用习惯和操作技能的基础上，加强学生利用大型分析仪器获得对物质组成和含量分析的综合知识的能力，提升分析和解决问题的能力，后期通过创新实验培养学生的开拓创新意识，为后续学习相关专业课程和开展科学研究奠定扎实的实践基础。

本书在编写过程中得到了编委所在院校的大力支持，汇集了安徽中医药大学、安徽医科大学、蚌埠医学院、皖南医学院、合肥学院、皖西学院、亳州学院等安徽省内高等医药院校专门从事仪器分析理论及实验教学的一线老师的智慧和辛勤付出！特别感谢安徽中医药大学谢晓梅教授在教材编写过程中给予的大力支持和无私奉献！本书的顺利出版，还得到了中国科学技术大学出版社的指导和支持,也得到了“省级仪器分析教学创新团队”的资助,在此一并感谢。

由于仪器设备种类、数量和实验教学时长等条件限制,以及编者的认识水平有限,书中难免存在不足之处,恳请广大师生在使用过程中提出宝贵意见,以便再版修订时进一步修改完善。

编　者

2023年2月

Preface

This book closely focuses on the requirements of the Ministry of Education of the People's Republic of China on the reform of experimental teaching in colleges and universities and the requirements of the construction of local high-level universities, which aims to improve the practical ability of students majoring in pharmacy and better promote the reform of teaching content and teaching methods. At the same time, it could strengthen students' practical application of English in their majors, cultivate high-level pharmaceutical talents with international competitiveness, and promote the exchange and learning of international students. In combination with the characteristics of higher medical colleges and the characteristics of the instrument analysis discipline, according to the requirements of the teaching syllabus of the instrument analysis experiment course for pharmaceutical majors, the editorial committee group combined with the current experimental teaching equipment and conditions, organized the preparation of the Instrument Analysis Experiment (Bilingual edition) textbook for the the instrument analysis experiment course for undergraduate majors such as pharmacy, Chinese pharmacy, pharmaceutical pharmaceutics and pharmaceutical engineering.

In terms of content, this book strives to fully absorb advanced experimental techniques. According to the inherent laws and links of instrument analysis methods, it is mainly about the use of analytical instruments, supplemented by its structure and principle, and focuses on the training of basic experimental operation ability. On the basis of cultivating good usage habits and operating skills of instrument specification, strengthen students' comprehensive knowledge of material composition and content analysis by using large analytical instruments, improve their ability to analyze and solve problems, and cultivate students' innovative awareness through innovation experiments in the later stage, laying a solid practical foundation for subsequent study of relevant professional courses and scientific research.

This book has been strongly supported by the colleges and universities that the editorial committee belong to during the preparation process， and has gathered all the wisdom and hard work of the front-line teachers specialized in the teaching of instrumental analysis theory and experiment from Anhui University of Chinese Medicine， Anhui Medical University， Bengbu Medical College， Wannan Medical College， Hefei University， West Anhui University， Bozhou University and other medical colleges and universities in Anhui Province. Special thanks to Professor Xiaomei Xie of Anhui University of Chinese Medicine for her support and selfless dedication in the preparation of textbook. At the same time， the successful publication of this book has also received the guidance and support of the University of Science and Technology of China Press. We would like to thank you.

Due to the limitations of the type and quantity of instruments， experimental teaching time and other conditions， as well as the limited level of understanding of the editors， there are inevitably shortcomings in the book. We sincerely invite teachers and students to put forward valuable suggestions during the use of the book， so as to further revise and improve it when republishing.

Editor

February 2023

目　录

第1章　仪器分析实验基本知识

1.1　仪器分析实验要求

仪器分析实验是学生在教师指导下进行的一种科学实践活动，让学生以分析仪器为工具并根据物质的物理性质或物理化学性质来获取物质的组成、含量、结构及相关信息。仪器分析实验已成为中药及医学类专业的重要课程之一。通过仪器分析实验课的学习，加强学生对分析仪器的基本原理、仪器构造及基本操作技能的掌握，并培养学生根据实验对象合理设计实验方案和解决实际问题的能力，最终养成严谨求实的科学态度、科技创新的精神和独立工作的能力。为了达到以上教学目标，教师和学生在实验课中必须遵循以下几点要求：

(1) 实验指导教师必须熟练掌握分析仪器及相关设备的使用与保养，充分理解实验内容，清楚实验过程的各个环节，及时告知学生分析仪器使用中应注意的事项，尽全力避免不规范操作对分析仪器和设备造成的损坏。

(2) 实验课前，学生需做好预习工作，明确实验的目的、方法、原理、操作程序及注意事项，特别是仪器的使用和安全事项。

(3) 实验过程中，学生需学会正确使用分析仪器，做好实验过程记录，并严格遵守实验室各项规章制度和管理措施。

学会正确使用分析仪器。学生在上课期间，需认真听取实验指导教师讲解仪器的使用方法。学生在实验过程中要严格按照仪器的操作规程操作仪器，认真练习操作技术，且必须做到先清楚操作程序才能动手。爱护实验仪器，切记不要随意摆弄和按动仪器的操作旋钮和按键。

做好实验过程记录。学生进入实验室要随身携带预先编好页码的实验预习记录本。实验过程中细心观察并详细记录实验现象、实验条件和测定的原始数据。应用钢笔或圆珠笔及时、完整、准确地将实验现象、实验条件和测定的原始数据记录于实验记录本上，绝不允许将以上信息先记在小纸条、手上或其他本子上再誊写到实验记录本上。原始记录是实验报告的一部分，尊重原始记录是基本的科学素养。如记录有误，可加注说明或将写错处用双线划去，在旁边写上正确数据；不得删改实验数据,否则无效。

严格遵守实验室各项规章制度和管理措施,服从教师及实验技术人员的指导。

(4) 实验结束后,学生应整理好仪器设备,清理实验台,经教师同意后,方可离开实验室。值日生应负责整理公用试剂,打扫卫生,并检查水、电、门窗是否关好等安全事宜。

(5) 实验课后,学生应认真写好实验报告,并按规定时间交给指导老师批改。实验报告的内容应包括实验名称、日期、目的、方法和原理、仪器及型号、所用试剂的规格和浓度、实验条件、实验操作步骤、实验数据(图)和实验现象、数据处理和结果分析以及问题讨论等。

Chapter 1　Basic Knowledge of Instrumental Analysis Experiments

1.1　Requirements for Instrumental Analysis

Instrumental analysis is a scientific practice activity carried out by students under the guidance of teachers, allowing students to use analytical instruments as a tool to obtain the composition, content, structure, and related information of substances according to their physical properties or physicochemical properties. Instrumental analysis has become one of the principle experiments courses in the traditional Chinese medicine and medicine majors. Through the study of the instrument analysis course, students can grasp the basic principles, instrument structure, and basic operation skills of analytical instruments. Thereby it can cultivate students' ability to rationally design experimental plans according to the experimental objects and to solve practical problems. Ultimately, it helps students develop a rigorous and realistic scientific attitude, the spirit of technological innovation, and the ability to work independently. To achieve the above teaching objectives, teachers and students must comply with the following requirements in the experimental class.

(1) Experiment instructors must be proficient in the use and maintenance of analytical instruments and related equipments. They should as well fully understand the content of the experiment, comprehend all aspects of the experimental process, and promptly inform students of the matters that should be paid attention to so as to avoid irregular operations when using the analytical instruments.

(2) Before the experiment, students need to preview the experimental content and clarify the purpose, method, principle, operation procedure, and precautions of the experiments especially the use of instruments and safety matters.

(3) During the experiment, students need to learn to use analytical instruments correctly, keep records of the experiment process, and strictly abide by the rules, regulations, and management measures of the laboratory.

Learn to use analytical instruments correctly. During the class, students need to listen carefully to the experimental instructor to explain how to use the instruments. During the experiment, students must first understand the operation procedures before starting, operate the instrument strictly in accordance with the operation procedures of the instrument, and practice the operation technique carefully. Students must take good care of the experimental instruments and remember not to fiddle with and press the operating knobs and buttons of the instrument.

Make a record of the experiment process. When entering the laboratory, students should carry a pre-numbered lab notebook. During the experiment, students should carefully observe and record the experimental phenomenon, conditions, and original data in detail. Students should use a pencil or ballpoint pen to record the experimental phenomena, conditions, and original data in the experimental record book timely, completely, and accurately. It is never allowed to record the above information on a small piece of paper, on your hand, or on other notebooks and then copy out the information on the experimental notebook. Please remember that the original record is part of the experimental report, and respecting the original record is the basis of scientific literacy. If there is an error in the record, you can add a description or cross out the wrong place with a double line, and write the correct data next to it. Please do not delete or modify the original experimental data, otherwise it will be invalid.

Strictly abide by the rules, regulations, and management measures of the laboratory, and obey the guidance of teachers and experimental technicians.

(4) After the experiment, students should arrange the equipment, clean the laboratory table, and leave the laboratory after the teacher consent. Students on duty should be responsible for arranging public reagents, cleaning, and checking safety issues such as water, electricity, doors and windows.

(5) After the experiment class, students should carefully write the experiment report and submit it to the instructor for correction within the prescribed time. The content of the experimental report should include the experimental name, date, purpose, method and principle, instrument model, specifications and concentrations of reagents used, experimental conditions, experimental operation steps, experimental data and experimental phenomena, data processing and result analysis, and question discussion etc.

1.2　仪器分析实验安全知识

在仪器分析实验中，经常使用各种药品和仪器设备，以及水、电、气，还会经常遇到高温、低温、高压、真空、高频和带有辐射源的实验条件和仪器。为确保实验过程中人身安全，学生进入实验室后必须严格遵守实验室的安全规则。

(1) 不得饮食、吸烟或把食具带进实验室。

(2) 要穿长袖且过膝的实验服，不准穿拖鞋和凉鞋，必须将长发束起或藏于帽内。

(3) 浓酸、浓碱具有很强的腐蚀性，容易对人体造成不同程度的伤害。使用时应注意以下几点：

防止浓酸、浓碱沾到皮肤或洒到衣服上。若浓硫酸洒到皮肤上，不要先用清水冲洗，而要迅速用布拭去，再用水冲洗。若其他酸沾到皮肤上，应立即用大量清水冲洗，再在皮肤上涂上3%～5%的碳酸氢钠溶液。若碱沾到皮肤上，应立即用大量水冲洗，再涂上硼酸溶液。需在通风橱中操作，且不允许直接加热。稀释浓硫酸时需在耐热容器中操作，且只能将浓硫酸慢慢注入水中，切不可将水倒入硫酸中。

(4) 使用易燃挥发性有机溶剂或有毒、刺激性气体(如 H_2S、HF、Cl_2、CO 等)必须在通风橱中进行，且不能俯向溶液闻挥发出的气体，用后应倒入回收瓶中，不准倒入下水道造成污染。

(5) 废液不得倒入下水道，需倒入指定的废液缸中。

(6) 若实验室着火，应先切断电源，再用干粉或气体灭火器灭火，不可直接泼水灭火，以防触电。

(7) 若有人触电，应先切断电源或用绝缘体将电线与人体分离，再实施抢救。

(8) 烫伤或烧伤时应及时处理，若受伤严重应立即送往医院进行抢救。

(9) 自觉养成节约用水的习惯，用完水应随手关紧水龙头。

(10) 严格按照规定操作使用气体钢瓶。

1.2 Safety Knowledge of Instrumental Analysis

During instrumental analysis experiments, we often use various drugs, instruments, water, electricity, gas, experimental conditions and instruments with high temperature, low temperature, high pressure, vacuum, high frequency, and radiation sources. To ensure personal safety during the experiment, students must strictly abide by the safety rules after entering the laboratory.

(1) Do not eat, drink, smoke or bring food utensils into the laboratory.

(2) Wear long-sleeved and knee-length lab coats. Slippers and sandals are not allowed. Long hair must be tied up or hidden in a hat.

(3) Concentrated acid and concentrated alkali are highly corrosive and can easily cause different degrees of damage to the human body. The following points should be paid attention to when the students use above reagents.

Prevent contact of concentrated sulfuric acid and alkali with the skin or spilling on clothes. If the concentrated sulfuric acid is spilled on the skin, do not rinse with water first, but quickly wipe it off with a cloth and then rinse with water. If other acids are spilled on the skin, rinse the skin immediately with plenty of water and then apply 3%～5% sodium bicarbonate solution on the skin. If the alkali is spilled on the skin, rinse immediately with plenty of water and then apply the boric acid solution on the skin.

Be sure to operate in a fume hood and do not heat the acid and alkali directly.

The dilution of concentrated sulfuric acid must be conducted in a heat-resistant container. Only the concentrated sulfuric acid can be slowly poured into the water, and water must not be poured into the sulfuric acid.

(4) The flammable volatile organic solvents or toxic and irritating gases (such as H_2S, HF, Cl_2, CO, etc.) must be used in a fume hood. Do not lean over to smell the volatilized gas. After being used, these solvents should not to be poured into the sewer and should be poured into the recycling bottle.

(5) The liquid waste should be poured into the designated liquid waste tank and should not be poured into the sewer.

(6) If there is a fire in the laboratory, cut off the power supply first and then use dry powder or gas fire extinguisher to put out the fire. Do not directly splash

water to extinguish the fire, in case of electric shock.

(7) If someone gets an electric shock, first cut off the power supply or separate the wires from the human body with an insulator and then rescue.

(8) The scald or burn should be dealt with in time, and in severe cases the injured should be sent to the hospital immediately.

(9) Consciously develop the habit of saving water and turn off the tap after use.

(10) Operate and use gas cylinders in strict accordance with regulations.

1.3 仪器分析实验数据记录与处理

1.3.1 实验记录

实验记录是对实验步骤、现象以及测量数据的及时、准确和清晰的记录,是追溯实验数据的直接证据。为保证实验数据的真实性,实验记录时应注意:

(1) 实验记录本应保持完整、不得缺页或挖补;如有缺页、漏页,应说明原因。

(2) 应使用钢笔、圆珠笔、签字笔等做实验记录,不得使用铅笔。

(3) 实验开始之前,记录实验时间、实验名称、实验目的、实验基本原理、仪器与试剂、实验内容和操作步骤等。

(4) 实验过程中,应及时、准确和清晰地记录实验中出现的各种现象以及测量数据。一定要本着实事求是和严谨的科学态度记录各种测量数据。切不可夹杂主观因素随意拼凑和伪造实验数据,也绝不允许将测量数据先记在小纸条、手上或其他本子上再誊写到实验记录本上。如果实验数据记录有误,可加注说明或将写错处用双线划去,在旁边写上正确数据;不得删改实验数据,否则无效。

(5) 实验过程中记录测量数据时,应严格按照所用仪器的精密度正确保留有效数字的位数。

(6) 实验过程中应记录所有的测量数据,即使有完全相同的数据,也要记录。

(7) 实验结束后,将实验记录本交给实验指导老师检查和签字。

1.3.2 数据的表达与处理

1. 数据的表达

做完实验后会获得大量数据,可用列表法将数据整齐有规律地表达出来。列表时应注意列表要有明确的名称,在表的每一行和每一列的第一栏注明详细的名称和单位,表中数据的有效数字的位数要合理。

2. 数据的取舍

为了衡量分析结果的精密度,会对同一批样品的重复测量数据 X_1,X_2,X_3,…,X_n 计算平均值。如果测定次数较多,可用标准偏差和相对标准偏差等表示结果的精密度。若某一数值偏差较大,则可以舍弃。

3. 作图

利用图形表达实验结果更直观,易显示出数据的特点,如极大值、极小值、转折点等。

1.3.3　实验报告

实验结束后，学生应使用专门的实验报告本及时写好实验报告，并对实验结果进行总结和讨论。实验报告一般包括以下内容：

(1) 实验名称、实验日期以及实验者。

(2) 实验目的：从掌握、熟悉以及了解三个层面简要说明本实验的目的和要求。

(3) 基本原理：用文字或图形简要说明实验的主要原理。

(4) 仪器与试剂：仪器的信息包括实验中使用的各种仪器的名称、型号、厂家以及仪器的参数；试剂的信息包括实验中使用的各种溶液和样品，并注明样品浓度等参数。

(5) 实验内容和操作步骤：简要列出各实验步骤。

(6) 数据记录和处理：采用表格、文字或图形列出测得的实验数据。根据实验要求计算实验结果，并给出结论。

(7) 讨论：对实验现象和实验结果的成功和失败原因进行讨论和分析，总结经验教训，提高分析和解决问题的能力。

(8) 思考题：完成实验指导教师布置的思考题。

1.3 Record and Treatment of Instrumental Analysis Experimental Data

1.3.1 Experimental Record

The experimental record is a timely, accurate and clear record of the experimental steps, phenomena and experimental data. It is the direct evidence to trace the experimental data. In order to ensure the authenticity of the experimental data, attention should be paid as follows when recording the experiment:

(1) The experimental records should be kept intact, and no pages should be missing or excavated; if there are missing pages, the reasons should be explained.

(2) Pen, ballpoint pens, signature pens, etc. should be used for experimental records, pencils are not allowed.

(3) Before the experiment, the students should record the time, name, purpose, basic principle, instruments and reagents, content, and operation steps of experiment.

(4) During the experiment, all kinds of phenomena and measurement data should be recorded timely, accurately and clearly. Be sure to record all kinds of experimental data in a realistic and rigorous scientific manner. The experiment date must not be pieced together or falsified with subjective factors, and the measurement data must not be written down on small slips of paper, hands or other notebooks and then transferred the experimental record book. If there is an error in the experimental data record, you can add a description or cross out the error with double lines, and write the correct data next to it. Do not delete or modify the experimental data, otherwise it will be invalid.

(5) When recording the data during the experiment, the number of significant digits should be correctly reserved in strict accordance with the precision of the instrument used.

(6) All measurement data should be recorded during the experiment, even if they are identical.

(7) After the experiment, hand over the experiment record to the experimental instructor for inspection and signature.

1.3.2 Expression and Processing of Experimental Data

1. Expression of Data

After the experiment, a large amount of data will be obtained, which can be expressed neatly and regularly by the method of lists. When making a list, it should be noted that the list should have a clear name, the detailed name and the units should be indicated in the first column of each row and column of the table, and the number of significant figures of the data in the table should be reasonable.

2. Selection of Data

In order to measure the precision of the analysis results, the average value of the repeated measurement data X_1, X_2, X_3, ..., X_n of the same batch of samples should be calculated. If the number of measurements is large, standard deviation and relative standard deviation can be used to express the precision of the results. If a certain value has a large deviation, it can be discarded.

3. Plotting

Graphical expression of experimental results is more intuitive, and it is easy to display the characteristics of the data, such as maximum value, minimum value, turning point, etc.

1.3.3 Experimental Report

After the experiment, students should use a special experimental report book to write the experiment report in time, and summarize and discuss the experimental results. The experimental report generally includes the following contents:

(1) The name of the experiment, the date of the experiment, and the experimenter.

(2) The purpose of the experiment: briefly explain the purpose and requirements of this experiment from three levels: mastery, familiarity, and understanding.

(3) Basic principle: briefly explain the main principle of the experiment with words or graphics.

(4) Instruments and reagents: the information of the instruments include the names, models, manufacturers, and instrument parameters of various instruments used in the experiment; the reagent information include various

solutions and samples used in the experiment, and parameters such as sample concentration should be indicated.

(5) Experimental content and operation steps: briefly list each experimental step.

(6) Data recording and processing: use tables, text or graphics to list the measured experimental data. Calculate the experimental results according to the experimental requirements, and provide conclusions.

(7) Discussion: discuss and analyze the reasons for the success or failure of experimental phenomena and results, summarize experiences, and improve the ability to analyze and solve problems.

(8) Questions: complete the questions assigned by the experimental instructor.

1.4　仪器分析实验室用水

1.4.1 分析实验室用水的规格

水是一种溶解能力很强的溶剂，有不少物质容易与水作用。天然水中含有许多杂质，经过净化处理后的城市自来水依旧难以满足实验室用水的要求。在化学实验，尤其是分析化学实验中，实验用水都必须先经过纯化。纯化的方法主要有蒸馏法、离子交换法、电渗析法和反渗透法。实验室用的纯水有蒸馏水、二次重蒸水和去离子水等。

根据中华人民共和国标准 GB 6682—92《分析实验室用水规格及试验方法》的规定，分析实验室用水分为三个级别：一级水，二级水和三级水，应符合表 1.1 所列要求。

表 1.1　分析实验室用水规格和要求

名称	一级	二级	三级
pH 范围(25 ℃)	—	—	5.0～7.5
电导率(25 ℃)/(ms/m)	≤0.01	≤0.01	≤0.50
可氧化物质含量(以 O 计)(mg/L)	—	≤0.08	≤0.40
吸光度(254 nm，1 cm 光程)	≤0.001	≤0.01	—
蒸发残渣(105 ℃ ± 2 ℃)含量/(mg/L)	—	≤1.0	≤2.0
可溶性硅(以 SiO_2 计)含量/(mg/L)	≤0.01	≤0.02	—

注 1：由于在一级水、二级水的纯度下，难以测定其真实的 pH 值，因此，对一级水、二级水的 pH 值范围不做规定。

注 2：由于在一级水的纯度下，难以测定可氧化物质和蒸发残渣，对其限量不做规定。可用其他条件和准备方法来保证一级水的质量。

一级水：用于有严格要求的分析实验，包括对颗粒有要求的实验，如高效液相色谱用水。一级水可用二级水经过石英设备蒸馏或离子交换混合处理后，再经过 0.2 μm 微孔滤膜过滤来制取。

二级水：用于无机痕量分析等实验，如原子吸收光谱分析用水。二级水可用多次蒸馏或离子交换等方法制取。

三级水：用于一般化学分析实验。三级水可用蒸馏或离子交换等方法制取。

通常,普通蒸馏水保存在玻璃容器中,去离子水保存在聚乙烯塑料容器中。用于痕量分析的高纯水则需要保存在石英或聚乙烯容器中。

1.4.2 各种纯度水的制备

蒸馏水:将自来水在蒸馏装置中加热汽化,然后将蒸汽冷凝即可得到蒸馏水。由于杂质离子一般不挥发,所以蒸馏水中所含杂质比自来水少得多,比较纯净,可达三级水的指标。

二次石英亚沸腾蒸馏水:为了获得比较纯净的蒸馏水,可以进行重蒸馏,并在准备重蒸馏的蒸馏水中加入适当的试剂以抑制某些杂质的挥发。如加入甘露醇能抑制硼的挥发,加入碱性高锰酸钾可以破坏有机物并防止二氧化碳蒸出。二次蒸馏水一般可达到二级水的指标。

去离子水:去离子水是用自来水或普通蒸馏水通过离子交换树脂后所得的水;纯度比蒸馏水纯度高,质量可达二级或一级水指标;但对非电解质及胶体物质无效,同时会有微量的有机物从树脂溶出。因此根据需要可将去离子水进行蒸馏以得到高纯水。市售70型离子交换纯水器可用于实验室制备去离子水。

在药物生产及药物制备中,水是用量最大、使用最广的一种原料。根据《中国药典》规定,制药用水的原水为自来水公司供应的自来水或沸开水,其质量必须符合国家生活饮用水卫生标准。原水不能直接用作制剂的制备或实验用水,必须经蒸馏法、离子交换法、反渗透法或其他适宜方法制得供药用的水,不含任何附加剂,不得有微生物污染。

1.4 Water Used in Analytical Laboratories

1.4.1 Specifications of Water Used in Analytical Laboratories

Water is a solvent with strong solubility, and many substances interact easily with water. Natural water contains many impurities, so the purified urban tap water still cannot meet the requirements of laboratory water. In chemical experiments, especially for analytical chemical experiments, water must first be purified. The purification methods mainly include distillation, ion exchange, electrodialysis and reverse osmosis. The pure water used in the laboratory includes distilled water, double-distilled water, and deionized water.

According to the provisions of the People's Republic of China standard GB 6682—92 *Water for Analytical Laboratory Specifications and Test Methods*, the water used in analytical laboratories is divided into three levels: first-grade water, second-grade water, and third-grade water, which should meet the requirements listed in Table 1.1.

Table 1.1 Specifications and requirements for water used in analytical laboratories

Item	First-grade	Second-grade	Third-grade
pH range (25 ℃)	—	—	5.0~7.5
Electrical conductivity(25 ℃)/(ms/m)	≤0.01	≤0.01	≤0.50
Oxidizable substance content (based on oxygen) (mg/L)	—	≤0.08	≤0.40
Absorbance (254 nm, 1 cm optical path)	≤0.001	≤0.01	—
Evaporation residue content(105 ℃ ± 2 ℃)/(mg/L)	—	≤1.0	≤2.0
Soluble silicon content (based on SiO_2)/(mg/L)	≤0.01	≤0.02	—

Note 1: Since it is difficult to determine the true pH value of first-grade and second-grade water, the pH range of first-grade and second-grade water is not specified.

Note 2: Because it is difficult to determine the oxidizable substances and evaporation residue of first-grade water, the limit is not specified. Other conditions and preparation methods can be used to ensure the quality of first-grade water.

First-grade water: It is used for analytical experiments with strict requirements, including the experiments with requirement for particles size, such as high-performance liquid chromatography. The first-grade water can be prepared by the second-grade water after the quartz equipment distillation or ion exchange mixing treatment, and then filtered by 0.2 μm microporous filter membrane.

Second-grade water: It is used for inorganic trace analysis and other experiments, such as atomic absorption spectrometry analysis. Second-grade water can be prepared by multiple distillation or ion exchange methods.

Third-grade water: It is used for general chemical analysis experiments. Third-grade water can be prepared by distillation or ion exchange.

Usually, ordinary distilled water is kept in glass containers, and deionized water is stored in polyethylene plastic containers. Highly pure water for trace analysis needs to be kept in quartz or polyethylene containers.

1.4.2 Preparation of Various Purities of Water

Distilled water: The tap water is heated and vaporized in a distillation device, and then the steam is condensed to obtain distilled water. Because impurity ions are generally involatile, distilled water contains much less impurities than tap water, which is relatively pure and can reach the index of tertiary water.

Secondary quartz sub-boiling distilled water: In order to obtain relatively pure distilled water, redistillation can be carried out, and appropriate reagents can be added into the distilled water ready for redistillation to inhibit the volatilization of certain impurities. For example, adding mannitol can inhibit the volatilization of boron, adding alkaline potassium permanganate can destroy organic matter and prevent the evaporation of carbon dioxide. Secondary distilled water can generally achieve the index of secondary-grade water.

Deionized water: Deionized water is the water obtained by passing tap water or ordinary distilled water through the ion exchange resin; its purity is higher than that of distilled water, and its quality can reach the index of the first-grade and secondary-grade water. However, this method is invalid for non-electrolyte and colloidal substances, and there will be trace organic matter dissolved from the resin. Therefore, deionized water can be distilled to obtain high-purity water. Commercially available Type 70 ion exchange water purifiers can be used to prepare deionized water in the laboratory.

During the production and preparation of drugs, water is the most widely used raw material. According to the provisions of *Chinese Pharmacopoeia*, the raw water for pharmaceutical water is tap water or boiling water supplied by a water company, and its quality must meet the national hygiene standards for drinking water. The raw water cannot be directly used for development of preparations or experimental water. It must be prepared by distillation, ion exchange, reverse osmosis or other suitable methods for medicinal purposes, without any additives and without microbial contamination.

1.5 仪器分析样品前处理技术

样品前处理，是指将样品分解，使被测组分定量地转入溶液中以便进行分析测定的过程。对于无机物可以采用溶解法、熔融法、烧结法或闭管法。对于有机物可以采用干灰化法、湿消化法或微波消解法。选用何种方法，取决于分析元素和被测样品的基本性质。本节主要介绍几种常用的方法。

1.5.1 干灰化法

样品通常先经 100～105 ℃干燥，除去水分及挥发物质，再将样品放入马弗炉中(通常为 450～550 ℃)被充分灰化。通常是将样品置于恒重的坩埚中，先在低温条件下缓慢灼烧，至完全灰化，冷却，用少量碱性或酸性物质(固定剂)湿润残渣，继续加热至酸性物质蒸汽逸尽。将坩埚连同残渣一同转入 450～550 ℃的高温炉中灼烧，放冷，称重。重复灼烧，直到恒重。最后留下不挥发的无机残留物。

这种方法能灰化大量的样品，操作者不需时常观测，操作简单，但容易导致被测成分元素损失，或可能与容器起反应，被氧化或被吸收，导致回收率低。

1.5.2 湿法消解

湿法消解是指用无机强酸和/或强氧化剂溶液将样品中的有机物质分解、氧化，使待测组分转化为可测定形态的方法。具体是在样品中加入氧化性酸或氧化剂，并同时加热消煮，使有机物质分解氧化成二氧化碳、水和各种气体。常用的氧化性酸和氧化剂有浓硝酸、浓硫酸、高氯酸、高锰酸钾、过氧化氢等。针对不同的样品选择不同的酸体系。针对 80 ℃以下的消解体系选用盐酸消解，针对 80～120 ℃的体系选择硝酸消解，针对 340 ℃左右的体系选用硫酸消解。混合酸的消解体系也较为常用，如盐酸-硝酸(适合 95～110 ℃的消解体系)，硝酸-高氯酸(适合 140～200 ℃的消解体系)，硝酸-硫酸(适合 120～200 ℃的消解体系)，硝酸-双氧水(适合 95～130 ℃的消解体系)。选择合适的酸体系对加快破坏有机物十分重要，同时要进行准确的温度控制，才能够达到理想的消解效果。

1.5.3　熔融法

熔融法是指将试样与酸性或碱性固体熔剂在坩埚中混合，并在 500～600 ℃高温下进行复分解反应，使待测组分转变为可溶于水或酸的化合物，如钠盐、钾盐、硫酸盐或氯化物等。不溶于水、酸或碱的无机试样一般可采用这种方法分解。使用的溶剂分为酸熔和碱熔两类。常见的酸性熔剂有焦硫酸钾和硫酸氢钾，适用于分解碱性或中性试样。常见的碱性熔剂有碳酸钠（或碳酸钾）、氢氧化钠（或氢氧化钾）、过氧化钠或其混合熔剂，适用于分解酸性试样，例如长石或酸性炉渣等。

1.5 Pretreatment Techniques for Instrumental Analysis of Samples

Sample pretreatment refers to the process of decomposing the sample and quantitatively transferring the measured components into the solution for analysis and determination. For inorganic substances, dissolution, melting, sintering, or closed-tube method can be used. For organic substances, dry ashing, wet digestion, or microwave digestion method can be used. The method chosen depends on the basic properties of the elements and the sample analyzed in the experiment. This section mainly introduces several commonly used methods.

1.5.1 Dry Ashing

The sample is generally dried at 100～105 ℃ firstly to remove moisture and volatile substances. Then the sample is placed in a muffle furnace (generally 450～550 ℃) to be fully ashed. Usually, the sample is placed in a crucible with constant weight, which is slowly burned at a low temperature until the sample is completely ashed, and the resulting sample is cooled down to the temperature. The residue is wetted with a small amount of alkaline or acidic substances (fixatives), and the residue is continued to be heated until the vapors of the acidic substances are exhausted. The crucible together with the residue are transferred to the furnace at 450～550 ℃ for burning. After burning, the crucible is put to be cool and weighed. The burning operation is repeated until constant weight. In the end, a non-volatile inorganic residue remains.

This method can ash a large number of samples, the operator does not need to observe frequently, and the operation is simple. But it is easy to cause the loss of the measured component elements, the reaction of the measured component elements with the container, or the oxidation or absorption of the measured component elements, resulting in the lower recovery rate.

1.5.2 Wet Digestion

Wet digestion refers to the method of decomposing and oxidizing the organic material in the sample with inorganic strong acid or strong oxidant solution, so that the components to be tested can be converted into a measurable form. Specifically, an oxidizing acid or an oxidizing agent is added to the sample, and the sample was heated at the same time to make the oxidized organic substances decompose and oxidize into carbon dioxide, water, and various gases. Commonly used oxidizing acids and oxidants are concentrated nitric acid, concentrated sulfuric acid, perchloric acid, potassium permanganate, hydrogen peroxide, etc. Different acid systems are selected for different samples. Hydrochloric acid is used for the digestion system below 80 ℃. Nitric acid is used for the digestion system at 80～120 ℃. Sulfuric acid is used for the digestion system at about 340 ℃. Mixed acid digestion systems are also commonly used, such as hydrochloric acid-nitric acid (suitable for digestion systems at 95～110 ℃), nitric acid-perchloric acid (suitable for digestion systems at 140～200 ℃), nitric acid-sulfuric acid (suitable for digestion systems at 120～200 ℃), nitric acid-hydrogen peroxide (suitable for digestion system at 95～130 ℃). It is very important to choose a suitable acid system to accelerate the destruction of organic matter. At the same time, accurate temperature control can achieve the ideal digestion effect.

1.5.3 Melting

The melting method refers to mixing the sample with an acidic or basic solid flux in a crucible, and performing a metathesis reaction at a high temperature of 500～600 ℃ to convert the components to be measured into compounds soluble in water or acid, such as sodium salts, potassium salt, sulfate or chloride, etc. Inorganic samples that are insoluble in water, acid or alkali can generally be decomposed by this method. The solvents used are divided into two categories: acid melting and alkali melting. Common acidic fluxes are potassium pyrosulfate and potassium hydrogen sulfate, which are suitable for decomposing alkaline or neutral samples. Common alkaline fluxes include sodium carbonate (or potassium), sodium hydroxide (or potassium), sodium peroxide or their mixed fluxes, which are suitable for decomposing acidic samples, such as feldspar or acidic slag.

第 2 章　验证性仪器分析实验

2.1　氟离子选择电极测定饮用水中的氟

2.1.1　实验目的

（1）掌握氟离子选择电极的使用方法。

（2）掌握直接电位法进行定量分析的原理和步骤。

（3）了解总离子强度调节缓冲溶液的构成及作用。

2.1.2　基本原理

1．氟离子选择电极

以 LaF_3 单晶作为敏感膜，封于塑料管的底部，敏感膜中常掺入少量的 EuF_3 以提高其导电性。在管内封装 0.1 mol/L NaF 和 0.1 mol/L NaCl 溶液，以银-氯化银电极（Ag-AgCl）为内参比电极构成氟离子选择电极。

进行 F^- 测定时，以氟离子选择电极为指示电极，饱和甘汞电极（SCE）为参比电极，两个电极插入待测溶液中构成测量电池。电池表示为

氟离子选择电极 | 溶液 | 饱和甘汞电极

在 25 ℃时，电池的电动势可以表示为

$$E = K - 0.059 \lg c_{F^-}$$

因此，在一定条件下，该电池电动势与溶液中的氟离子浓度的对数呈线性关系。

2．测定的影响因素

常见阴离子 NO_3^-、SO_4^{2-}、PO_4^{3-}、Cl^-、Br^-、I^-、HCO_3^- 对 F^- 测定不产生干扰，但是 OH^- 会对氟离子的测定产生影响，主要是由于在敏感膜表面会发生如下反应：

$$LaF_3 + 3OH^- \longequal La(OH)_3 + 3F^-$$

F^-为电极本身影响造成的干扰。而在pH过低时，易形成HF，HF_2^- 等，降低氟离子活度，因此需要控制溶液的pH在5.0～5.5之间。

可以加入电解质氯化钠控制离子强度。此外，F^-易与Al^{3+}、Fe^{3+}等离子形成配合物，使得氟离子浓度降低。因此在测定时需要加入配合能力强的配位剂，如柠檬酸盐，消除Al^{3+}、Fe^{3+}等离子的干扰，提高测定的准确性。

综上，为了保证测定的准确度，需要向标准溶液和待测溶液中加入总离子强度调节缓冲溶液（TISAB）以控制溶液pH，维持溶液中离子强度保持恒定并消除共存离子的干扰。

通过离子计也可以对氟离子浓度进行直接测定，测定方法与测定溶液pH方法相似，需保持标准溶液和水样中离子强度基本相同。

2.1.3　仪器与试剂

1. 仪器

821型离子计，氟离子选择电极，饱和甘汞电极，电磁搅拌器。

2. 试剂

氟离子标准贮备液（1.0×10^{-1} mol/L），TISAB溶液。

2.1.4　实验步骤

1. 氟离子选择电极的预处理

在使用之前，将氟离子选择电极置于10^{-3} mol/L NaF溶液中活化1～2 h，使用时，用去离子水冲洗干净。然后，与饱和甘汞电极组成测量电池，测定纯水的电动势。当测定的纯水电动势读数在+300 mV以上说明电极活化完成；若小于+300 mV，可更换数次蒸馏水直至读数在+300 mV以上。

2. 溶液的配制

（1）氟离子标准贮备液的配制

将NaF于120 ℃烘干2 h，称取4.199 g溶于1000 mL容量瓶中，用蒸馏水稀释定容，摇匀，得1.0×10^{-1} mol/L氟离子标准贮备液。

（2）TISAB溶液的配制

在1000 mL烧杯中加入500 mL去离子水，称取58 g氯化钠，12 g柠檬酸钠加入其中，随后加入57 mL冰醋酸，搅拌至完全溶解。缓缓加入6 mol/L NaOH溶液调节溶液pH至5.0～5.5之间，冷却至室温后，转移至1000 mL容量瓶中定容摇匀。

(3) 氟离子标准溶液的配制

用移液管移取 10 mL 1.0×10^{-1} mol/L 氟离子标准贮备液于 100 mL 容量瓶中,加入 10 mL TISAB 溶液,用去离子水定容,获得 1.0×10^{-2} mol/L 氟离子溶液。依次类推,采用逐级稀释法获得 1.0×10^{-3}～1.0×10^{-6} mol/L 氟离子标准溶液。

(4) 试液的配制

取饮用水 50 mL 于 100 mL 容量瓶中,加入 10 mL TISAB 溶液,用去离子水定容。

3. 氟离子的测定

(1) 标准曲线法

将用移液管移取不同浓度的标准溶液 20 mL 分别置于 50 mL 烧杯中,在搅拌条件下依次测定各标准溶液的电位值,平行测定三次。以电位 E 值为纵坐标,样品浓度的对数值为横坐标,绘制 $E-\lg c_{F^-}$ 标准曲线。

移取试样溶液 25 mL 于 50 mL 烧杯中,与标准溶液在相同条件下测定试样的电位 E 值,平行测定三次。

(2) 标准加入法

在试样中加入 1 mL 1.0×10^{-3} mol/L 氟离子标准溶液,测定加入前后的电位值,平行测定三次。

2.1.5 数据记录与处理

1. 数据记录

将实验数据记入表 2.1。

表 2.1

c_{F^-} (mol/L)	1.0×10^{-2}	1.0×10^{-3}	1.0×10^{-4}	1.0×10^{-5}	1.0×10^{-6}	试样
E(mV)						
$\lg c_{F^-}$						

2. 结果计算

(1) 以电位 E 值为纵坐标,样品浓度的对数值为横坐标,绘制 E-$\lg c_{F^-}$ 标准曲线。

(2) 建立线性回归方程和相关系数,计算饮用水中氟离子含量。

2.1.6　注意事项

（1）仪器使用前需要开机预热。

（2）通常由稀至浓分别测定标准溶液。

（3）测定一系列标准溶液后，要将电极清洗至原空白电位值，随后测定试样溶液。

（4）测定电动势时，氟离子选择电极和饱和甘汞电极在溶液中的高度要适中，既要在液面以下，又要避免搅拌子碰到，造成电极的损坏。

（5）测定过程中，搅拌速度要适中。

（6）氟离子选择电极的晶片上如有油污，可用酒精和丙酮轻轻擦拭，随后用蒸馏水清洗干净。

2.1.7　思考题

（1）总离子强度调节剂的作用是什么？

（2）在使用氟离子选择电极进行测量前，需要经过怎样处理，达到什么要求？

（3）比较标准曲线法和标准加入法的测定结果。

Chapter 2 Confirmatory Instrumental Analysis Experiments

2.1 Determination of Fluoride Ions in Drinking Water by Fluoride Ion Selective Electrode

2.1.1 Objectives

(1) Master the use of fluoride ion selective electrodes.

(2) Master the principles and steps of direct potential method for quantitative analysis.

(3) Understand the composition and function of ionic strength buffer solutions.

2.1.2 Principles

1. Fluoride Ion Selective Electrode

The LaF_3 single crystal is used as the sensitive film, which is sealed at the bottom of the plastic tube. A small amount of EuF_3 is often doped into the sensitive film to improve its conductivity. 0.1 mol/L NaF and 0.1 mol/L NaCl solution is encapsulated in the tube, and a silver-silver chloride electrode (Ag-AgCl) is used as the internal reference electrode to form the fluoride ion selective electrode.

For the determination of F^-, the fluoride ion selective electrode is used as the indicator electrode, and the saturated calomel electrode (SCE) is used as the reference electrode, and the two electrodes are inserted into the solution to be tested to form a measurement cell. The cell is represented as:

Fluoride Ion Selective Electrodes | Solutions | Saturated Calomel Electrodes

At 25 ℃, the electromotive force of the cell can be expressed as follow:

$$E = K - 0.059\ \lg c_{F^-}$$

Therefore, under certain conditions, the electromotive force of the cell is linear with the logarithm of the fluoride ion concentration in the solution.

2. The Influencing Factors of the Measurement

Common anions NO_3^-, SO_4^{2-}, PO_4^{3-}, Cl^-, Br^-, I^-, HCO_3^- do not interfere with the determination of F^-, but OH^- will affect the determination of fluoride ions. The following reactions could occur on the membrane surface:

$$LaF_3 + 3OH^- \longequal La(OH)_3 + 3F^-$$

F^- is the interference caused by the influence of the electrode itself. When the pH value is too low, it is easy to form HF, HF_2, etc., which reduces the activity of fluoride ions. Therefore, it is necessary to control the pH of the solution between 5.0～5.5.

The electrolyte sodium chloride can be added to control the ionic strength. In addition, F^- easily forms complexes with Al^{3+}, Fe^{3+} and other ions, which reduces the concentration of fluoride ions. Therefore, it is necessary to add a complexing agent with strong coordination ability, such as citrate, to eliminate the interference of Al^{3+}, Fe^{3+} plasma and improve the accuracy of the determination.

In summary, in order to ensure the accuracy of the determination, it is necessary to add total ionic strength adjustment buffer solution (TISAB) to the standard solution and the solution being tested to control the pH of the solution, maintain the ionic strength in the solution constant and eliminate the interference of coexisting ions.

The fluoride ion concentration can also be directly measured by an ion meter. The measurement method is similar to the method for measuring the pH of the solution, and the ionic strength in the standard solution and the water sample should basically be kept the same.

2.1.3 Apparatus and Reagents

1. Apparatus

821 ion meter, fluoride ion selective electrode, saturated calomel electrode, electromagnetic stirrer.

2. Reagents

Fluoride ion standard stock solution(1.0×10^{-1} mol/L), TISAB solution.

2.1.4 Procedures

1. Pretreatment of Fluoride Ion Selective Electrodes

Before use, the fluoride ion selective electrode is activated in a 1.0×10^{-3} mol/L NaF solution for 1 to 2 h. Then, the fluoride ion selective electrode is rinsed with deionized water and forms a measuring cell with a saturated calomel electrode to measure the electromotive force of pure water. When the measured pure water electromotive force reading is above +300 mV, the electrode activation is completed; if it is less than +300 mV, the distilled water can be replaced several times until the reading is above +300 mV.

2. Preparation of the Solution

(1) Preparation of Fluoride Ion Standard Stock Solution

The NaF should be dried at 120 ℃ for 2 h. Then, 4.199 g of NaF is dissolve it in a 1000 mL volumetric flask, diluted with distilled water, and shaken to obtain 1.0×10^{-1} mol/L fluoride ion standard stock solution.

(2) Preparation of TISAB Solution

500 mL of deionized water is added into a 1000 mL beaker. Then, 58 g of sodium chloride and 12 g of sodium citrate are added into beaker. Next, 57 mL of glacial acetic acid is added into the above solution and stirred to make a uniform solution. 6 mol/L NaOH solution is used to adjust the pH of the solution between 5.0 and 5.5. After cooling to room temperature, the solution is transferred into a 1000 mL volumetric flask and shaken up to constant volume.

(3) Preparation of Fluoride Ion Standard Solution

Transfer 10 mL of 1.0×10^{-1} mol/L fluoride ions standard stock solution into a 100 mL volumetric flask by pipette. Then, add 10 mL of TISAB solution and dilute with deionized water to obtain 1.0×10^{-1} mol/L fluoride ion solution. By analogy, the stepwise dilution method is used to obtain 1.0×10^{-3} ~1.0×10^{-6} mol/L fluoride ion standard solution.

(4) Preparation of Test Solution

Add 50 mL of drinking water to 100 mL volumetric flask and mixed with 10 mL of TISAB solution. Then, the solution is diluted to 100 mL with deionized water.

3. Determination of Fluoride Ions

(1) Standard Curve Method

20 mL of standard solutions of different concentrations were respectively

moved into 50 mL beakers by pipette. The potential value of each standard solution should be measured sequentially under stirring conditions 3 times in parallel. Taking the potential E as the ordinate and the logarithm of the sample concentration as the abscissa, the standard curve about E and $\lg c_{F^-}$ was obtained.

Add 25 mL of the sample solution into a 50 mL beaker by pipette. The value of potential E of the sample should be measured under the same conditions as the standard solution 3 times in parallel.

(2) Standard Addition Method

Add 1 mL of 1.0×10^{-3} mol/L fluoride ion standard solution to the tested sample. The potential value before and after adding should be recorded 3 times in parallel.

2.1.5 Data Recording and Processing

1. Data Recording

Record the experimental data in Table 2.1。

Table 2.1

c_{F^-} (mol/L)	1.0×10^{-2}	1.0×10^{-3}	1.0×10^{-4}	1.0×10^{-5}	1.0×10^{-6}	Tested sample
E (mV)						
$\lg c_{F^-}$						

2. Result Calculation

(1) Taking the potential E as the ordinate and the logarithm of the sample concentration as the abscissa, draw the standard curve of E-$\lg c_{F^-}$.

(2) Establish a linear regression equation and correlation coefficient to calculate the fluoride ion content in drinking water.

2.1.6 Cautions

(1) The instrument needs to be turned on and warmed up before use.

(2) Standard solutions are usually determined separately from dilute to be concentrated.

(3) After measuring a series of standard solutions, the electrode should be cleaned to the original blank potential value, and then the sample solution

should be measured.

(4) When measuring the electromotive force, the height of the fluoride ion selective electrode and the saturated calomel electrode in the solution should be moderate, not only below the liquid level, but also to avoid the stirring bar touching and damaging the electrode.

(5) During the measurement, the stirring speed should be moderate.

(6) If there is oil on the wafer of the fluoride ion selective electrode, it can be gently wiped with alcohol and acetone, and then cleaned with distilled water.

2.1.7 Questions

(1) What is the role of total ionic strength modifiers?

(2) Before using the fluoride ion selective electrode for measurement, what should be done and what are the requirements?

(3) Compare the determination results of standard curve method and standard addition method.

2.2　醋酸电位滴定

2.2.1　实验目的

(1) 掌握酸度计(pH 计)的使用方法。
(2) 掌握电位滴定法终点确定的方法。
(3) 掌握实验有关数据处理。
(4) 比较酸碱指示剂滴定法和酸碱电位滴定法的异同。

2.2.2　基本原理

醋酸 HAc($Ka=1.8\times10^{-5}$,$pKa=4.74$)是弱酸,可以用 NaOH 标准溶液测定 HAc 的浓度。

$$HAc + NaOH = NaAc + H_2O$$

在滴定过程中,随着滴定剂 NaOH 的不断加入,HAc 与 NaOH 溶液发生定量反应,溶液的 pH 不断在变化,接近化学计量点时发生"滴定突跃"。因此测量溶液 pH 的变化,就能确定滴定终点。滴定过程中,每加一次滴定剂,测一次 pH,在接近化学计量点时,每次滴定剂加入量为 0.05 mL,滴定到 HAc 溶液 pH 接近 12 时,停止滴定。这样就得到一系列滴定剂用量(NaOH 体积)V 和相应的 pH,并绘制 pH-V 滴定曲线,对此数据进行一阶微分和二阶微分的换算,可确定 HAc 的滴定终点,从而确定 HAc 的准确浓度。因此采用电位滴定法可测定 HAc 浓度。

2.2.3　仪器与试剂

1. 仪器

酸度计一台、电磁搅拌器、复合玻璃电极、移液管。

2. 试剂

HAc、NaOH 标准溶液、pH = 4.00 和 pH = 6.86 的标准缓冲溶液。

2.2.4　实验步骤

酸度计的使用方法如下:

(1) 仪器安装:电极安装(电极浸泡至少 4 小时,新电极浸泡至少 24 小时)。
(2) 接通电源,仪器预热约 30 分钟。

(3) pH 校正:

① 选择开关置“pH”档,“斜率”旋钮顺时针旋到底,“温度”旋钮置此标准溶液的温度。

② 电极用蒸馏水洗净,吸干,放入 pH = 6.86 的标准缓冲溶液中,用“定位”旋钮调节数字为 6.86。

③ 电极用蒸馏水洗净,吸干,放入 pH = 4.00 的标准缓冲溶液中,用“斜率”旋钮调节数字为 4.00(该步骤所选择的标准缓冲溶液 pH 应该与被测溶液 pH 接近,即酸性溶液选择酸性标准缓冲溶液,碱性溶液选择碱性标准缓冲溶液)。

(4) 滴定:精密移取 10.00 mL 的 HAc 标准溶液于 50 mL 小烧杯中,放入搅拌子,加入复合玻璃电极并确认溶液浸没末端球形玻璃膜。开启电磁搅拌器,用 NaOH 标准溶液进行滴定,记录消耗的 NaOH 标准溶液的体积 V 和测定醋酸溶液的 pH。按下述方式进行滴定。

第一次粗略滴定:每次消耗约 1.00 mL NaOH 标准溶液,记录消耗的 NaOH 标准溶液的准确体积 V 和溶液的 pH(用于确定滴定突跃范围)。

第二次精确滴定:从 V_{NaOH}标准溶液体积为 0.00 mL 时开始滴定,在突跃范围前和突跃范围后,按照每次消耗约 1.00 mL NaOH 标准溶液进行滴定;在突跃范围内,按照每次消耗约 0.05 mL NaOH 标准溶液进行滴定。每次滴定要求记录消耗的 NaOH 标准溶液的体积 V 和溶液的 pH(用于数据处理,作图)。

2.2.5 数据记录与处理

(1) 绘制 pH-V 滴定曲线。以滴定剂 NaOH 标准溶液的体积 V 为横坐标,HAc 溶液的 pH 为纵坐标作图。用“三切法”作图确定滴定终点,计算 HAc 浓度。

(2) 绘制 $\Delta pH/\Delta V$-V 滴定曲线。以滴定剂 NaOH 标准溶液的体积 V 为横坐标,以 $\Delta pH/\Delta V$ 为纵坐标,绘制一阶微分曲线,$\Delta pH/\Delta V$ 极大值对应的 V 为滴定终点。

(3) 绘制 $\Delta^2 pH/\Delta V^2$-V 滴定曲线。以滴定剂 NaOH 标准溶液的体积 V 为横坐标,以 $\Delta^2 pH/\Delta V^2$ 为纵坐标,绘制二阶微分曲线,$\Delta^2 pH/\Delta V^2$ 由极大正值到极大负值与横坐标相交处对应的 V 为滴定终点。

2.2.6 思考题

(1) 与指示剂法测定滴定终点相比,使用电位滴定法测定滴定终点有哪些优点和缺点?

(2) 当 HAc 被 NaOH 完全中和时,反应结束时的 HAc 溶液的 pH 是否等于 7? 为什么?

2.2 Potentiometric Titration of Acetic Acid

2.2.1 Objectives

(1) Master the use of acidity meter (pH meter).

(2) Master the method regarding determination of the end point of potentiometric titration.

(3) Master the experimental data processing.

(4) Compare the similarities and differences between acid-base indicator titration and acid-base potential titration.

2.2.2 Principles

HAc ($Ka=1.8\times10^{-5}$, $pKa=4.74$) is a weak acid, and the concentration of HAc can be determined with NaOH standard solution.

$$HAc + NaOH \Longrightarrow NaAc + H_2O$$

During the titration process, the titrant NaOH should be continually added. HAc reacts quantitatively with NaOH, the pH of the solution is constantly changed, and a "titration abrupt" occurs when the stoichiometric point is reached. Therefore, the end point of the titration can be determined by measuring the change in pH of the solution. During the titration process, the pH should be measured after each time the titrant is added. When the pH approaches the stoichiometric point, the amount of titrant added each time should be as small as 0.05 mL. The titration process can be stopped after the pH exceeds the stoichiometric point. In this way, a series of titrant dosages (NaOH volume) V and the corresponding pH value are obtained, and a pH-V titration curve can be drawn. From the pH-V titration curve, the titration point of HAc can be determined, so that the HAc concentration in the mixture can be calculated. Therefore, the HAc concentration can be determined by potentiometric titration.

2.2.3 Instruments and Reagents

1. Instruments

Acidity meter, electromagnetic stirrer, composite glass electrode, pipette.

2. Reagents

HAc solution, NaOH standard solution, standard buffer solution of pH equal to 4.00 and 6.86.

2.2.4 Procedures

How to use the acidity meter:

(1) Instrument installation: electrode installation (electrode soaking for at least 4 hours, new electrode for at least 24 hours).

(2) Turn on the instrument power to warm up for about 30 minutes.

(3) Calibration of the pH meter:

Set the selector switch to "pH", turn the "slope" knob clockwise to the end, and set the "temperature" knob to the temperature of the standard solution.

The electrode must be washed with distilled water, sucked dry, and placed in a standard buffer solution with pH equal to 6.86. The pH value should be adjusted to 6.86 with the "positioning" knob.

Wash the electrode the electrode with distilled water, suck dry, and put it in a standard buffer solution with pH equal to 4.00. Adjust the pH value to 4.00 with the "slope" knob. (The pH of the standard buffer solution selected in this step should be close to the pH of the solution being tested, that is, the acidic standard buffer solution is selected for the acidic solution being tested, and the alkaline standard buffer solution is selected for the alkaline solution being tested.)

(4) Titration:

Put 10.00 mL of HAc into a 50 mL small beaker, put a stir bar, and then immerse the composite electrode into the solution. Turn on the electromagnetic stirrer, titrate the solution with NaOH standard solution, and measure the corresponding pH value.

Rough titration: record the corresponding pH value after the titration of every 1.00 mL NaOH solution (for determining the titration jump range).

Precise titration: before and after the jump range, measure the corresponding pH after the titration of every 1.00 mL; within the jump range, measure the corresponding pH after the titration of every 0.05 mL (for data processing and graphing).

2.2.5 Data Recording and Processing

(1) Plot pH-V titration curve. The volume V of the standard solution of titrator NaOH was taken as the horizontal coordinate and the pH value of acetic acid solution was taken as the vertical coordinate. Use the "three-cut method" to plot the titration endpoint and calculate the HAc concentration.

(2) Plot the $\Delta pH/\Delta V$-V titration curve. Plot the first order differential curve with the volume V of the standard solution by taking titration agent NaOH as the horizontal coordinate and $\Delta pH/\Delta V$ as the vertical coordinate, and the V corresponding to the maximum value of $\Delta pH/\Delta V$ is the end point of titration.

(3) Plot the $\Delta^2 pH/\Delta V^2$-V titration curve. Taking the volume V of the standard solution of titrator NaOH as the horizontal coordinate and $\Delta^2 pH/\Delta V^2$ as the vertical coordinate, plot a second-order differential curve. The V at the intersection of the zero line of the vertical coordinate and $\Delta^2 pH/\Delta V^2$ from the maximum positive value to the maximum negative value is the end point of the titration.

2.2.6 Questions

(1) What are the advantages and disadvantages of using potentiometric titration method compared to the indicator method to determine the titration end point?

(2) Is the pH at the end of the reaction equal to 7 when HAc is completely neutralized by NaOH? Why?

2.3 邻二氮菲分光光度法测定水中微量铁

2.3.1 实验目的

(1) 掌握分光光度计的操作使用。
(2) 熟悉标准曲线法进行定量分析的原理和程序。
(3) 了解显色反应在测定水中微量铁含量的原理和方法。

2.3.2 基本原理

铁是水中一种常见的杂质,对饮水中微量铁进行检查和测定常利用比色法。

当铁以 Fe^{3+} 离子形式存在于溶液中时,可预先用还原剂(盐酸羟胺或维生素 C 等)将其还原为 Fe^{2+} 离子:

$$2Fe^{3+} + 2NH_2OH \cdot HCl \longrightarrow 2Fe^{2+} + N_2\uparrow + 2H_2O + 4H^+ + 2Cl^-$$

亚铁离子与邻二氮菲生成稳定的橙红色配合物离子:

$$Fe^{2+} + 3\,\text{(phen)} \rightleftharpoons [Fe(\text{phen})_3]^{2+}$$

显色时溶液 pH 值应为 2～9,若酸度过高(pH<2),则显色缓慢而色浅;若酸度过低,则二价铁离子容易水解。

根据上述反应,采用比色法并根据 Beer 定律,在最大吸收波长 508 nm 测定标准溶液的吸光度 A,绘制标准曲线或建立回归方程即可测定水溶液中溶解的总铁含量。

2.3.3 仪器与试剂

1. 仪器

UV1000 紫外可见分光光度计,容量瓶(50 mL、100 mL),吸量管(1 mL、2 mL、5 mL),移液管(10 mL、20 mL),量筒(100 mL),洗耳球,擦镜纸及大张

滤纸。

2. 试剂

铁标准溶液(100 μg/mL),邻二氮菲水溶液(0.15%),盐酸羟胺水溶液(10%,用时新配),NaAc溶液(1.0 mol/L),NaOH溶液(0.1 mol/L),HCl溶液(6.0 mol/L)。

2.3.4 实验步骤

1. 可见分光光度计的操作规程

见附录7。

2. 铁标准溶液的制备

准确称取0.8643 g分析纯级$NH_4Fe(SO_4)_2 \cdot 12H_2O$置于200 mL烧杯中,加入20 mL 6 mol/L HCl溶液和少量水,溶解后转移至1 L容量瓶中,加水定容至刻度,摇匀。

3. 标准曲线的绘制

用移液管移取10 mL浓度为100 μg/mL铁标准溶液于100 mL容量瓶中,加入2 mL HCl溶液,用水定容至刻度,摇匀。此溶液含有的Fe^{3+}浓度为10 μg/mL。

准备6个50 mL容量瓶,用吸量管分别移取0.0 mL、2.0 mL、4.0 mL、6.0 mL、8.0 mL、10.0 mL 10.0 μg/mL铁标准溶液于6个50 mL容量瓶中,每个容量瓶再分别加入1 mL的盐酸羟胺水溶液、2 mL的邻二氮菲水溶液、5 mL的NaAc溶液,用水稀释至刻度后摇匀,放置10 min后待测。用1 cm比色皿,以试剂空白(即不加标准铁液)为参比,在所选波长下,测量各溶液的吸光度A。以铁浓度c为横坐标,吸光度A为纵坐标,绘制标准曲线,同时建立回归方程。

4. 水样测定

以自来水为样品,准确吸取澄清水样5.00 mL(或适量),置于50 mL容量瓶中。按上述制备标准曲线的方法配制溶液并测定吸光度A,根据测得的吸光度A求出水中总铁量。

2.3.5 数据记录与处理

1. 数据记录

将实验数据记入表2.2。

表 2.2

加入标准溶液的体积(mL)	2.0	4.0	6.0	8.0	10.0	水样
配制的标准溶液浓度(μg/mL)						
吸光度(A)						

2. 结果计算

(1) 绘制标准曲线图。

(2) 求出回归方程、相关系数。

(3) 由标准曲线图计算水样中铁的浓度。

2.3.6 注意事项

(1) 要遵守实验平行原则,空白与标准系列均应按相同的操作步骤进行操作,包括加试剂的量、顺序、时间等应一致。

(2) 绘制标准曲线图要用坐标纸,建立回归方程的同时考察相关系数。

(3) 计算水样中铁的浓度时不要遗漏稀释倍数。

2.3.7 思考题

(1) 显色反应操作中,各标准溶液与样品溶液的含酸量不同,对显色有无影响?

(2) 根据标准曲线测得的数据,计算浓度与吸光度之间的线性相关系数,判断线性关系。

(3) 作图法和计算法所得结果有无显著性差异? 分析原因。

2.3 Spectrophotometric Determination of Trace Iron in Water by 1,10-Phenanthroline

2.3.1 Objectives

(1) Master the operation and use of the spectrophotometer.

(2) Be familiar with the principles and procedures of quantitative analysis by standard curve method.

(3) Understand the principles and methods of color reaction used to determine the trace iron content in water.

2.3.2 Principles

Iron is a common impurity in water, and colorimetry is often used to inspect and determine the trace iron in drinking water.

When the iron exists in the solution in the form of Fe^{3+} ions, it can be reduced to Fe^{2+} ions with a reducing agent (hydroxylamine hydrochloride or vitamin C, etc.) in advance:

$$2Fe^{3+} + 2NH_2OH \cdot HCl \longrightarrow 2Fe^{2+} + N_2\uparrow + 2H_2O + 4H^+ + 2Cl^-$$

Ferrous ion can combine with 1,10-phenanthroline monohydrate to form a stable orange-red complex ion:

$$Fe^{2+} + 3\,\text{phen} \rightleftharpoons [Fe(\text{phen})_3]^{2+}$$

The pH value of the solution should be controlled at 2～9 to make the orange-red of the complex ion. If the acidity is too high (pH $<$ 2), the color change of the complex ion is slow and the color is light; if the acidity is too low, the ferrous ions will be hydrolyzed.

According to the above two reactions, the absorbance A of the standard solution should be measured at the maximum absorption wavelength of 508 nm by the colorimetric method and Beer's law. The total iron content dissolved in the aqueous solution can be determined by drawing a standard curve or establishing a regression equation.

2.3.3 Instruments and Reagents

1. Apparatus

UV-Vis spectrophotometer, volumetric flasks (50 mL, 100 mL), pipettes (1 mL, 2 mL, 5 mL), suction pipets (10 mL, 20 mL), graduated cylinder (100 mL), washing ear balls, lens cleaning paper, and large sheets of filter paper.

2. Reagents

Iron standard solution (100 μg/mL), 0.15% aqueous solution of 1,10-phenanthroline monohydrate, 10% aqueous solution of hydroxylamine hydrochloride (newly prepared), 1 mol/L NaAc solution, 0.1 mol/L NaOH solution, and 6 mol/L HCl solution.

2.3.4 Procedures

1. Operation Procedures for Visible Spectrophotometers

See Appendix 7.

2. Preparation of Iron Standard Solutions

0.8643 g of pure AR grade $NH_4Fe(SO_4)_2 \cdot 12H_2O$ is added into a 200 mL beaker, then 20 mL of 6 mol/L HCl solution and a small amount of water are added into above solution. After dissolved, the mixture is transferred to a 1 L volumetric flask, diluted with water to the mark, and shaken well.

3. Preparation of Standard Curve

10 mL of 100 μg/mL iron standard solution and 2 mL of HCl solution are added into a 100 mL volumetric flask using a suction pipet. Then, the solution is diluted to the mark with water and is stirred evenly. The concentration of Fe^{3+} contained in this solution is 10 μg/mL.

0.0 mL, 2.0 mL, 4.0 mL, 6.0 mL, 8.0 mL, 10.0 mL of 10.0 μg/mL iron standard solution are added into six 50 mL volumetric flasks by pipette, respectively. Then, 1 mL of 10% aqueous solution of hydroxylamine hydrochloride, 2 mL of 0.15% aqueous solution of 1, 10-phenanthrene monohydrate, and 5 mL of NaAc solution, are added into the above solution and diluted with water to the mark. The mixtures must then be set aside for 10 min. The absorbance of each solution at the selected wavelength is measured with a 1 cm cuvette. The blank reagent without standard iron solution is used as a reference. Taking the iron concentration as the abscissa and the absorbance A as the ordinate, a standard curve and a regression equation can be established at the same time.

4. Determination of Water Samples

Tap water is used as the sample, 5.00 mL of the clarified water sample is taken and placed in a 50 mL volumetric flask. The absorbance A of the sample is measured the according to the method for preparing the standard curve above. The total iron content in the water can be calculated according to the measured absorbance A.

2.3.5 Data Recording and Processing

1. Data Recording

Record the experimental data in Table 2.2.

Table 2.2

Volume of standard solution added (mL)	2.0	4.0	6.0	8.0	10.0	Water sample
Concentration of the prepared standard solution (μg/mL)						
Absorbance (A)						

2. Result Calculation

(1) Plot the standard curve.

(2) Find the regression equation, correlation coefficient.

(3) Calculate the concentration of Fe ions from the standard curve.

2.3.6 Cautions

(1) Comply with the principle of experimental parallelism. For example, the blank and standard series should be operated according to the same operation steps, including the amount, sequence, and the time of adding reagents.

(2) Graph paper should be used to draw the standard curve. The regression

equation should be established and the correlation coefficient is examined.

(3) Do not omit the dilution ratio when calculating the iron concentration in the water sample.

2.3.7 Questions

(1) In the color reaction operation, if the acid content between standard solution and the sample solution is different, does it affect the color reaction?

(2) According to the data obtained by preparing the standard curve, please determine the linear correlation coefficient between concentration and absorbance to determine the linear relationship.

(3) Are there significant differences between the results obtained by the drawing method and the calculation method? Analyze the reasons.

2.4　喹啉的荧光特性和含量测定

2.4.1　实验目的

(1) 掌握荧光分光光度计的工作原理。
(2) 熟悉喹啉激发光谱和荧光光谱的绘制方法。
(3) 了解溶液 pH 和卤化物对喹啉荧光的影响。

2.4.2　基本原理

喹啉在稀酸溶液中是强的荧光物质,有 250 nm 和 350 nm 两个激发波长,荧光发射峰在 450 nm。利用其荧光光谱和激发光谱可以对喹啉进行定性定量分析。

2.4.3　仪器与试剂

1. 仪器

荧光分光光度计和石英吸收池,容量瓶(50 mL、250 mL、100 mL),吸量管(10 mL),擦镜纸。

2. 试剂

喹啉贮备液(100 μg/mL),喹啉标准溶液(10.0 μg/mL,将喹啉贮备液稀释10 倍,即得),NaBr 溶液 0.05 mol/L,缓冲溶液(pH 为 1.0、2.0、3.0、4.0、5.0、6.0),H_2SO_4 溶液(0.05 mol/L),硫酸喹啉片。

2.4.4　实验步骤

1. 标准溶液配制

取 50 mL 容量瓶,加入喹啉标准溶液 4.00 mL,用 0.05 mol/L H_2SO_4 溶液稀释至刻度,摇匀。

2. 样品溶液配制

取 4～5 片硫酸喹啉片称重,在研钵中研磨,准确称取 0.1 g,用 1 mol/L H_2SO_4 溶液溶解,并用去离子水定容至 100 mL,摇匀。

取上述溶液 5.0 mL 置于 50 mL 容量瓶中,用 0.05 mol/L H_2SO_4 溶液稀释至刻度,摇匀。

3．绘制激发光谱和荧光光谱

以 $\lambda_{em}=450$ nm，在 200～400 nm 范围扫描标准溶液激发光谱；以 $\lambda_{ex}=250$ nm 或 350 nm，在 400～600 nm 范围分别扫描标准溶液的荧光光谱。

4．标准溶液和样品溶液测定

将激发波长固定在 250 nm 或 350 nm，测量发射波长为 450 nm 下溶液的荧光强度，记录标准溶液和样品溶液的荧光强度。

5．pH 对喹啉溶液荧光强度的影响

取 6 只 50 mL 容量瓶，分别加入喹啉标准溶液 4.00 mL，并分别用 pH 为 1.0、2.0、3.0、4.0、5.0、6.0 的缓冲溶液稀释至刻度，摇匀。分别测定 6 种溶液的荧光强度。

6．卤化物对喹啉溶液荧光强度的影响

取 5 只 50 mL 容量瓶，分别加入喹啉标准溶液 4.00 mL，并分别加入 0.05 mol/L NaBr 溶液 1.0 mL、2.0 mL、4.0 mL、8.0 mL、16.0 mL，用 0.05 mol/L H_2SO_4 溶液稀释至刻度，摇匀。测定 5 种溶液的荧光强度。

2.4.5 数据记录与处理

(1) 样品溶液测定：

标准溶液的荧光强度(F)：________________；样品溶液的荧光强度(F)：________________。

(2) pH 与荧光强度的关系(表 2.3)。

表 2.3

pH	1.0	2.0	3.0	4.0	5.0	6.0
荧光强度(F)						

(3) 卤化物对喹啉溶液荧光强度的影响(表 2.4)。

表 2.4

NaBr 溶液体积(mL)	1.0	2.0	4.0	8.0	16.0
荧光强度(F)					

2.4.6　注意事项

喹啉溶液必须新鲜配制并避光保存。

2.4.7　思考题

（1）为什么测量溶液荧光时比色皿放置的位置必须和激发光的方向成直角？

（2）本实验中能用 0.05 mol/L HCl 溶液来代替 0.05 mol/L H_2SO_4 溶液吗？为什么？

2.4 Fluorescence Properties and Content Determination of Quinoline

2.4.1 Objectives

(1) Master the principle of fluorescence spectrophotometer.

(2) Be familiar with the method of mapping the fluorescence spectrum and excitation spectrum of quinoline.

(3) Understand the effect of solution pH and halide on quinoline fluorescence.

2.4.2 Principles

Quinoline is a strong fluorescent substance in dilute acid solution, and it has two excitation wavelengths, 250 nm and 350 nm. Its fluorescence emission peak is at 450 nm. Quinoline can be qualitatively and quantitatively analyzed by its fluorescence spectrum and excitation spectrum.

2.4.3 Instruments and Reagents

1. Instruments

Fluorescence spectrophotometer and quartz cuvette, volumetric flask (50 mL, 250 mL, 1000 mL), pipette (10 mL), lens cleaning paper.

2. Reagents

Quinoline stock solution (100 μg/mL: weigh 120.7 mg of quinoline sulfate dihydrate accurately, add 50 mL of 1 mol/L H_2SO_4 to dissolve, dilute it in a 1 L volumetric flask with deionized water, and shake well), the quinoline standard solution (10.0 μg/mL: dilute the quinoline stock solution 10 times to obtain the quinoline standard solution), NaBr solution (0.05 mol/L), buffer solution (pH 1.0, 2.0, 3.0, 4.0, 5.0, 6.0), H_2SO_4 solution (0.05 mol/L), quinoline sulfate tablets.

2.4.4 Procedures

1. Preparation of Standard Solutions

Add 4.00 mL of quinoline standard solution into a 50 mL volumetric flask, dilute the solution to the mark with 0.05 mol/L H_2SO_4 solution, and shake well.

2. Preparation of Sample Solutions

Weigh 4～5 quinoline sulfate tablets, grind them in a mortar, and then weigh 0.1 g of the sample powder accurately. Then dissolve the sample powder with 1 mol/L H_2SO_4 to 100 mL in a volumetric flask, and shake the solution well.

Take 5.00 mL of the above solution in a 50 mL volumetric flask, dilute the solution to the mark with 0.05 mol/L H_2SO_4, and shake well.

3. Plotting of Excitation Spectra and Fluorescence Spectra

Scan the emission spectrum of the standard solution in the range of 200～400 nm with λ_{em} equal to 450 nm. Scan the fluorescence spectrum of the standard solution in the range of 400～600 nm with λ_{ex} equal to 250 nm and 350 nm, respectively.

4. Determination of Sample Solutions and Standard Solutions

Fix the excitation wavelength at 250 nm or 350 nm, measure the fluorescence intensity of the solution at the emission wavelength of 450 nm, and record the fluorescence intensity of the standard solutions and sample solutions.

5. Relationship between the pH and Fluorescence Intensity

Take six 50 mL volumetric flask and add 4.00 mL of quinoline standard solution respectively, dilute them to the mark with buffer solutions with pH of 1.0, 2.0, 3.0, 4.0, 5.0, 6.0, respectively, and shake well. Measure the fluorescence intensities of the six solutions.

6. Effect of Halide on Fluorescence Intensity of Quinoline Solution

Take five 50 mL volumetric flasks, add 4.00 mL of quinoline standard solution and 1.0 mL, 2.0 mL, 4.0 mL, 8.0 mL, and 16.0 mL of 0.05 mol/L NaBr solution, respectively, dilute to the mark with 0.05 mol/L H_2SO_4 solution, and shake well. Measure the fluorescence intensities of the above five solutions.

2.4.5 Data Recording and Processing

(1) Sample solution determination:

Fluorescence intensity of standard solution (F): ________; fluorescence intensity of sample solution (F): ________.

(2) Relationship between the pH and fluorescence intensity (Table 2.3).

Table 2.3

pH	1.0	2.0	3.0	4.0	5.0	6.0
Fluorescence intensity (F)						

(3) Effect of halide on fluorescence intensity of quinoline solution (Table 2.4).

Table 2.4

Volume of NaBr solution (mL)	1.0	2.0	4.0	8.0	16.0
Fluorescence intensity (F)					

2.4.6 Cautions

Quinoline solution must be prepared before useing and stored away from light.

2.4.7 Questions

(1) Why should the cuvette be placed at a right angle to the direction of the excited light when measuring the fluorescence of the solution?

(2) Can 0.05 mol/L H_2SO_4 solution be replaced with 0.05 mol/L HCl solution in this experiment? Why?

2.5　阿司匹林红外吸收光谱的测定

2.5.1　实验目的

（1）掌握红外吸收光谱的基本原理。
（2）掌握红外光谱固体样品的制备方法。
（3）熟悉红外光谱仪的基本结构。

2.5.2　基本原理

红外吸收光谱是指物体分子受到红外辐射后产生的吸收光谱，其原理是波长为 2.5～25 μm（波数 4000～400 cm^{-1}）的中红外光与分子相互作用时，如果红外辐射的频率等于分子中某个基团的振动频率，并且该振动形式引起了分子偶极矩的变化，那么该基团吸收该频率的红外光，并产生振动能级的跃迁。红外吸收光谱是以波数 σ（单位：cm^{-1}）为横坐标，透光率 T 为纵坐标而绘制的曲线。根据红外吸收光谱中特征吸收峰的位置、数目、相对强度和形状（峰宽）等参数，它可以用于已知化合物的定性鉴别和未知化合物的结构分析。该方法常用于中药化学成分的结构分析。

2.5.3　仪器与试剂

1. 仪器
FT-IR-650 型傅里叶变换红外光谱仪，红外专用压片机，压片模具，玛瑙研钵。
2. 试剂
溴化钾粉末（光谱纯），无水乙醇（A. R.），阿司匹林（A. R.）。

2.5.4　实验步骤

（1）按照 FT-IR-650 型傅里叶变换红外光谱仪操作规程，打开红外光谱仪并稳定 15 min 以上。同时进入对应的计算机工作站，打开操作软件，进行参数设置。

（2）取烘干的 100～200 mg 溴化钾粉末，在玛瑙研钵中充分磨细（颗粒约 2 μm），使之混合均匀。取出约 80 mg 粉末均匀铺洒在干净的压模内，于压片机上制成透明薄片。将制备好的样品插入红外光谱仪的样品池处，从 4000～400 cm^{-1}

波数进行扫描,作为扫描背景。

(3) 取 1～2 mg 阿司匹林,加入烘干的 100～200 mg 溴化钾粉末,按照上述步骤进行同样操作,得到阿司匹林的红外吸收光谱。

(4) 结束实验,关闭工作站和红外光谱仪。

2.5.5 数据记录与处理

按照阿司匹林的结构对红外光谱图中的主要吸收峰进行归属。

2.5.6 注意事项

(1) 试样纯度应在 98%以上,如纯度太低则会给解析图谱带来困难,有时会造成误判,事先应尽量采用各种分离手段来制纯样品,样品应干燥。

(2) KBr 粉末或结晶在 110 ℃的烘箱中干燥 2～3 h,然后放在干燥器中保管。

(3) 在压片制样过程中,物料必须磨细并混合均匀,加入模具中需均匀平整,否则不易获得透明均匀的薄片。溴化钾极易受潮,因此制样操作应在低湿度环境中或在红外灯下进行。

2.5.7 思考题

(1) 傅里叶变换红外光谱仪是基于什么原理进行分光的?与色散型红外分光计相比在性能上有何特点?

(2) 在压片操作中应注意什么?

2.5 Determination of Infrared Absorption Spectrum of Aspirin

2.5.1 Objectives

(1) Master the principles of infrared absorption spectrum.

(2) Master the preparation method of solid sample for infrared absorption spectrum.

(3) Be familiar with the basic structure of the infrared spectrometer.

2.5.2 Principles

Infrared absorption spectroscopy refers to that when the middle infrared light with wavelength of 2.5～25 μm (wave number 4000～400 cm^{-1}) reacts with the molecules, the frequency of infrared radiation is equal to the vibration frequency of a molecular group, and the vibration form causes the change of molecular dipole moment, then the group absorbs the frequency of infrared light, and the vibration level transition. Infrared absorption spectrum is a curve with wave number σ (unit: cm^{-1}) as the horizontal coordinate and transmittance T as the vertical coordinate. The position, number, relative intensity, and shape (peak width) of characteristic absorption peaks in infrared absorption spectrum can be used for qualitative identification of known compounds and structural analysis of unknown compounds. This method is often used for structural analysis of chemical constituents in traditional Chinese medicine.

2.5.3 Instruments and Reagents

1. Apparatus

FT-IR-650 Fourier transform Infrared spectrometer, tablet press (with special mold), tablet mold, agate mortar.

2. Reagents

Potassium bromide (spectrum pure), anhydrous ethanol (A. R.), aspirin (A. R.).

2.5.4 Procedures

(1) According to the FT-IR-650 Fourier transform infrared spectrometer operating procedures, turn on the infrared spectrometer and stabilize for more than 15 min. At the same time, enter the corresponding computer workstation, open the operating software, and set the parameters.

(2) Take the dried 100～200 mg potassium bromide powder and grind well in an agate mortar (particles about 2 μm) to mix well. Then take out about 80 mg powder evenly spread in a clean press mold and make transparent flakes on a tablet press. Insert the prepared sample into the sample pool of the infrared spectrometer and scan from 4000～400 cm^{-1} wave number as the background for the scanning.

(3) Take 1～2 mg of aspirin, add 100～200 mg of dried potassium bromide (KBr) powder, follow the above steps to obtain the infrared absorption spectrum of aspirin.

(4) After the experiment, shut down the workstation and infrared spectrometer.

2.5.5 Data Recording and Processing

Locate the main absorption peaks in the infrared spectra according to the structure of aspirin.

2.5.6 Cautions

(1) The purity of the sample should be above 98%, if the purity is too low, it will bring difficulties to analyze spectrum and sometimes cause misjudgment. In advance, various separation means should be used to prepare pure samples, which should be dried.

(2) Dry KBr powder or crystals in an oven at 110 ℃ for 2～3 h, and then store in the dry chamber.

(3) In the process of tablet preparation, the material must be ground and mixed evenly, and it needs to be uniform and flat when added to the mold, otherwise it would not be easy to obtain a transparent and uniform flake. Potassium bromide is extremely susceptible to moisture, so sample preparation

should be carried out in a low humidity environment or under infrared lamps.

2.5.7 Questions

(1) What principles does the Fourier transform infrared spectrometer (FT-IR) adopt for spectroscopy? What are the differences of the performance of FT-IR compared to dispersive infrared spectrometers?

(2) What are the critical steps in the tablet pressing operation?

2.6　原子吸收法测定感冒冲剂中的铜

2.6.1　实验目的

(1) 掌握中药中微量金属元素的定量分析方法。

(2) 熟悉火焰法测量条件的选择。

(3) 了解原子吸收仪的操作使用。

2.6.2　基本原理

当物质分子引入原子吸收分光光度计原子化装置中时,物质分子转变成基态气态原子;当辐射源提供的能量与原子由基态跃迁到激发态所需能量相同时,气态原子中的外层电子就吸收该能量的辐射而产生跃迁;与分子吸收测定相类似。所测吸光度与该元素原子浓度之间符合 Beer 定律:$A = Kc$(K 是随实验条件而变的常数)。采用标准加入法对元素进行定量分析。

2.6.3　仪器与试剂

1. 仪器

(1) 原子吸收分光光度计,铜空心阴极灯,乙炔气。

(2) 石英亚沸高纯水蒸馏器。

2. 试剂

(1) 铜标准贮备液:将光谱纯试剂铜溶于一定浓度优级纯 HNO_3 中,制得 2 mg/mL 铜标准贮备液。

(2) 铜标准液:将铜标准贮备液用超纯水稀释至 20 μg/mL 即可。

(3) 感冒冲剂。

2.6.4　实验步骤

(1) 测量条件:

波长:324.7 nm;光谱带宽:0.45 nm;灯电流:2 mA;燃烧高度:6 mm;空气和乙炔流量:6.5∶1.0。

(2) 仪器操作。

(3) 取样品适量,精密称定,溶解后制成样品溶液。

(4) 标准曲线的绘制和铜含量的测定。分别精密量取铜标准液 0 mL、0.125 mL、0.25 mL、0.50 mL、0.75 mL、1.0 mL 于 10 mL 容量瓶中;精密加入上述样品溶液 8 mL,用 1% HNO_3 稀释至刻度,按上述仪器工作条件分别测定溶液的吸光度(A 值)。绘制 A-c 标准曲线,建立回归方程。

(5) 由标准曲线或回归方程获得感冒冲剂中铜的浓度,并计算其百分含量。

2.6.5　数据记录与处理

1. 数据记录

将实验数据记入表 2.5。

表 2.5

	0	1	2	3	4	5	样品
标准溶液浓度(μg/mL)							
吸光度(A)							

2. 数据处理

(1) 绘制标准曲线,建立回归方程和找到线性相关系数。

(2) 计算铜的含量。

2.6.6　注意事项

(1) 在使用标准曲线法绘制标准曲线时,每份样品溶液的量要准确相等。

(2) 实验完毕后,用超纯水清洗管路和雾化器喷嘴 3 min。

(3) 先关乙炔气,后关空气。

2.6.7　思考题

(1) 说明本实验的主要干扰因素及其消除措施。

(2) 比较两种方式获得结果是否存在差异,分析原因。

2.6 Determination of Copper in Cold Granules by Atomic Absorption Spectrometry

2.6.1 Objectives

(1) Master the quantitative analysis methods for trace metal elements in traditional Chinese medicine.

(2) Be familiar with the selection of measurement conditions for the flame method.

(3) Learn how to operate an atomic absorption spectrometer.

2.6.2 Principles

When a molecular substance is introduced into the atomization device of the atomic absorption spectrophotometer, the molecule is transformed into ground-state gaseous atoms. When the energy provided by the radiation source is equivalent to the energy required for the transition of the atom from the ground state to the excited state, the outer electrons of the gaseous atom will absorb the radiation of the energy and then make a transition, which is similar to molecular absorption spectrometry. The relationship between the measured absorbance and the concentration of the element's atoms follows Beer's law: $A = Kc$ (K is a constant that varies with experimental conditions). Quantitative analysis of the elements is performed by using the standard addition method.

2.6.3 Instruments and Reagents

1. Instruments

(1) Atomic absorption spectrophotometer, copper hollow cathode lamp, acetylene gas.

(2) Quartz second boiling high pure water distiller.

2. Reagents

(1) Copper standard stock solution: dissolve spectrum pure reagent copper in a certain concentration of guaranteed reagent HNO_3 to prepare 2 mg/mL

copper standard stock solution.

(2) Copper standard solution: dilute the stock solution with ultrapure water to 20 μg/mL.

(3) Cold granules.

2.6.4 Procedures

(1) Measurement conditions:

Wavelength: 324.7 nm; spectral bandwidth: 0.45 nm; lamp current: 2 mA; burning height: 6 mm; air ∶ acetylene flow: 6.5 ∶ 1.0.

(2) Instrument operation.

(3) Take an appropriate amount of sample and weigh it accurately. Then dissolve the sample to make the sample solution.

(4) Drawing of standard curve and determination of copper content. Precisely measure 0 mL, 0.125 mL, 0.25 mL, 0.50 mL, 0.75 mL, and 1.0 mL of the copper standard solution, respectively, and place them into 10 mL volumetric flasks. Add precisely 8 mL of the above-mentioned sample solution, dilute it to the mark with 1% HNO_3 (at the same time as the blank reagent), and measure the absorbance (the value of A) under the above-mentioned working conditions of the instrument. Plot the A-c standard curves and establish the regression equations.

(5) Find the concentration of cold granules from the standard curves or regression equations, then calculate its content.

2.6.5 Data Recording and Processing

1. Data Recording

Record the experimental date in Table 2.5。

Table 2.5

	0	1	2	3	4	5	sample
Concentration of Standard Solution (μg/mL)							
Absorbance (A)							

2. Data Processing

(1) Plot a standard curve, and establish a regression equation and find the linear correlation coefficient.

(2) Calculate the copper content.

2.6.6 Cautions

(1) When the standard addition method is used to plot the standard curves, the amount of each sample solution should be exactly equal.

(2) After the experiment, rinse the pipeline and atomizer nozzle with ultrapure water for 3 min.

(3) Turn off acetylene gas first, then air.

2.6.7 Questions

(1) Explain the main interference factors of this experiment and the elimination measures.

(2) Compare whether there are differences in the results obtained by the two methods and analyze the reasons.

2.7 核磁共振氢谱法确定有机化合物的分子结构

2.7.1 实验目的

（1）了解核磁共振波谱法的基本原理和傅里叶变换核磁共振仪的工作原理。
（2）掌握 Bruker AVANCE NEO 600 MHz 的操作技术。
（3）熟练掌握液体核磁共振波谱仪的制样技术。
（4）学会用^{1}H-NMR 谱图鉴定有机化合物的结构。

2.7.2 基本原理

^{1}H-NMR 的基本原理遵循的是核磁共振波谱法的基本原理。核磁共振波谱法直接获取的信息是化学位移。由于受到诱导效应、磁各向异性效应、共轭效应、范德华效应、浓度、温度以及溶剂效应等因素的影响，化合物中各种基团都有各自的化学位移值的范围，因此可以根据化学位移值判断谱峰的归属。^{1}H-NMR 中各峰的面积比与所含的氢的原子个数成正比，因此可以推断各基团所对应的氢原子的相对数目，还可以作为核磁共振波谱法定量分析的依据。偶合常数与峰形也是核磁共振波谱法可以得到的两个重要信息。它们可以提供分子内各基团之间的位置和相互连接的信息。根据以上的信息和已知的化合物分子式就可推出化合物的分子结构。

2.7.3 仪器与试剂

1. 仪器

Bruker AVANCE NEO 600 MHz 型核磁共振波谱仪（瑞士 Bruker 公司），直径为 5 mm 的核磁共振标准样品管（1 个），滴管（1 个）。

2. 试剂

TMS（四甲基硅烷），$CDCl_3$（氘代氯仿），分子式为 $C_8H_{10}O_2$ 的未知样品。

2.7.4 实验步骤

1. 配制样品

在核磁共振标准样品管中加入 2 mg 待测样品，并加入 0.5 mL 氘代氯仿及 1～2 滴TMS(内标)，盖上样品管盖，轻轻摇匀，待样品完全溶解后测试。

2. 测谱

(1) 样品管外部用天然真丝布擦拭干净后再插入转子中，放在深度规中量好高度。严格按照操作规程(此处操作失误有可能摔碎样品管并损害探头)。按下“Lift on/off”键，此键灯亮。当听到计算机有一声鸣叫时，仪器弹出原有的样品管。若听到孔穴中发出气流向上喷射的呼呼声，表明无样品管，此时等待探头穴中向上的气流可以托住样品管时，方可将样品管放到探头穴口，放入样品管。立即再按一下“Lift on/off”键，灯熄灭，样品管慢慢落到待测位置。

(2) 将仪器调节到可作常规氢谱的工作状态。

(3) 建立一个新的实验数据文件。

(4) 锁场。

(5) 调匀场。

(6) 设置采样参数。

(7) 自动设置接收机增益。

(8) 开始采样。

(9) 进行傅里叶变换。

(10) 调相位。

(11) 标定作为标准峰的化学位移值。

(12) 根据需要对选定的峰进行积分。

(13) 标出所需峰的化学位移值。

(14) 做打印前的准备工作。

(15) 打印。

2.7.5 数据记录与处理

对^{1}H-NMR 谱图中各信号峰进行归属，并推断分子式 $C_8H_{10}O_2$ 对应的化学结构。

2.7.6　注意事项

（1）样品浓度不宜太大。

（2）根据氢谱和具体样品的要求设定参数。

（3）在测量样品管高度时，要求做到准确无误。

（4）把样品放入探头时，一定要严格按照操作规程进行。

（5）不能随意修改实验参数，尤其不能改动功率参数。

（6）一定要仔细调好仪器的分辨率。

2.7.7　思考题

（1）一幅^{1}H-NMR谱图能提供哪些参数？每个参数是如何与分子结构相联系的？

（2）什么是自旋-自旋耦合的 $n+1$ 规律？如何应用 $n+1$ 规律解析谱图？

（3）说明化学位移与偶合常数之间的关系。

2.7 Identification of the Molecular Structure of Organic Compounds by ^{1}H-NMR Spectroscopy

2.7.1 Objectives

(1) Understand the basic principle of nuclear magnetic resonance spectroscopy and the working principle of Fourier transform nuclear magnetic resonance instrument.

(2) Master the operation method of Bruker AVANCE NEO 600 MHz.

(3) Achieve proficiency in the sample preparation technology of liquid NMR spectrometer.

(4) Learn to identify structures of organic compounds using ^{1}H-NMR spectrum.

2.7.2 Principles

The basic principles of ^{1}H-NMR follows the basic principles of nuclear magnetic resonance spectroscopy. The primary information directly obtained by NMR spectroscopy is the chemical shift. Due to the influence of factors such as induction effect, magnetic anisotropy effect, conjugation effect, van der Waals effect, concentration, temperature and solvent effect, various groups in the compound have their own chemical shift values. The chemical shift value determines the group to which the peak belongs. The area ratio of each peak in ^{1}H-NMR is proportional to the number of hydrogen atoms contained, so the relative number of hydrogen atoms corresponding to each group can be inferred, which can also be used as the basis for quantitative analysis of nuclear magnetic resonance spectroscopy. Coupling constants and peak shapes are the other two important information obtained by NMR spectroscopy. They can provide information on the positions and interconnections between groups within the molecule. According to the above information and the known molecular formula of the compound, the molecular structure of the compound can be deduced.

2.7.3 Instruments and Reagents

1. Instruments

Bruker AVANCE NEO 600 MHz nuclear magnetic resonance spectrometer (Bruker company, Switzerland), nuclear magnetic resonance standard sample tube ($\Phi = 5$mm), dropper.

2. Reagents

TMS(tetramethylsilane); $CDCl_3$(deuterated chloroform); unknown sample (molecular formula: $C_8H_{10}O_2$).

2.7.4 Procedures

1. Prepare the Sample Solution

Add 2 mg of the sample to the NMR standard sample tube, then add 0.5 mL of deuterated chloroform and 1～2 drops of TMS (as internal standard) in turn. Cover the sample tube, shake it gently until the sample is completely dissolved.

2. Spectrum Measurement

(1) Wipe the outside of the sample tube with a natural silk cloth, then insert it into the rotor, and place it in the depth gauge to measure the height. Strictly follow the operation procedures (Please avoid breaking the sample tube, which may damage the probe). When pressing the "Lift on/off" button, the button light will be on. When you hear a beep from the computer, it means that the instrument has ejected the original sample tube. If you hear the whirring sound of air jetting upward in the hole, it indicates that there is no sample tube. At this time, wait for the upward air flow in the probe hole to hold the sample tube, and then put the sample tube into the probe hole. Press the "Lift on/off" button immediately, the botton light will be off, and the sample tube will slowly fall into testing place.

(2) Adjust the instrument to a working state that can be used for conventional hydrogen spectroscopy.

(3) Create a new experimental data file.

(4) Lock field.

(5) Mix field.

(6) Set the test parameters.

(7) Automatically set the receiver gain.

(8) Start to test.

(9) Perform Fourier transform.

(10) Modulate phase.

(11) Calibrate the chemical shift value of the standard peak.

(12) Integrate the selected peaks as needed.

(13) Indicate the chemical shift value of the desired peak.

(14) Do the preparatory work before printing.

(15) Print.

2.7.5 Data Recording and Processing

Attribute the signal peaks in ^{1}H-NMR spectra and deduce the chemical structure corresponding to the molecular formula of $C_8H_{10}O_2$.

2.7.6 Cautions

(1) The sample concentration should not be too large.

(2) Set parameters according to the requirements of hydrogen spectrum and specific samples.

(3) When measuring the height of the sample tube, it is required to be accurate.

(4) When placing the sample into the probe, it must be carried out in strict accordance with the operating procedures.

(5) The experimental parameters cannot be modified at will, especially the power parameters.

(6) Be sure to adjust the resolution of the instrument carefully.

2.7.7 Questions

(1) What information can a ^{1}H-NMR spectrum provide? How does each parameter relate to the molecular structure?

(2) What is the $n+1$ law of spin-spin coupling? How can the $n+1$ rule be applied to analyze the spectrum?

(3) Explain the relationship between chemical shift and coupling constant.

2.8　薄层色谱法鉴别复方磺胺甲噁唑片

2.8.1　实验目的

（1）掌握物质比移值的计算方法及薄层板色谱法鉴别物质的方法。

（2）熟悉薄层色谱的基本操作。

（3）了解薄层色谱法在药物鉴别中的应用。

2.8.2　基本原理

利用固定相吸附、流动相解吸附和混合物各组分的性质差异，利用薄层色谱法对复方磺酸甲噁唑片中的磺胺甲恶唑和甲氧苄啶进行分离。在紫外光照射下，因组分吸收紫外光导致出现暗斑，从而实现不同组分的检视。计算得到不同组分的 R_f 值，并与对照品的进行比较，从而实现复方磺酸甲噁唑片的鉴别。

2.8.3　仪器与试剂

1．仪器

紫外分析仪，硅胶（GF_{254}），层析缸，研钵，量筒（25 mL、500 mL），锥形瓶（10 mL），漏斗（9 cm），滤纸（9 cm），点样毛细管。

2．试剂

三氯甲烷，甲醇，二甲基甲酰胺，磺胺甲噁唑对照品，甲氧苄啶对照品，复方磺胺甲噁唑片（每片中含磺胺甲噁唑应为0.360～0.440 g，含甲氧苄啶应为72.0～88.0 mg）。

2.8.4　实验步骤

1．硅胶 GF_{254} 薄层板的制备

称取硅胶 GF_{254} 10 g 置于研钵中，加入 20 mL 0.7%的羧甲基纤维素钠溶液，向同一方向研匀，倾倒在玻璃板上使其成薄层后，置水平台面上晾干，备用。

2．对照品溶液的制备

（1）取磺胺甲噁唑对照品 0.2 g，加甲醇 10 mL 溶解，作为磺胺甲噁唑对照品溶液。

（2）取甲氧苄啶对照品 40 mg，加甲醇 10 mL 溶解，作为甲氧苄啶对照品溶液。

(3) 取磺胺甲噁唑对照品 0.2 g 和甲氧苄啶对照品 40 mg,加甲醇 10 mL 溶解,作为混合对照品溶液。

3. 供试品溶液的制备

取复方磺胺甲噁唑片细粉适量(约相当于磺胺甲噁唑 0.2 g),加甲醇 10 mL 溶解,振摇,滤过,取滤液作为供试品溶液。

4. 薄层色谱试验

吸取上述各溶液 5 μL,分别点在同一硅胶 GF_{254} 板上,以三氯甲烷:甲醇:二甲基甲酰胺(20:2:1)为展开剂,展开,晾干,置紫外灯(波长 254 nm)下检视。供试品溶液所显两种成分的主要斑点的位置和颜色应与对照品溶液的斑点相同。

2.8.5 数据记录与处理

将实验数据记入表 2.6。

表 2.6

	对照品溶液		混合对照品		供试品溶液	
	磺胺甲噁唑	甲氧苄啶	磺胺甲噁唑	甲氧苄啶	斑点 A	斑点 B
原点至斑点中心的距离 (mm)						
原点至溶液前沿的距离 (mm)						
R_f						
结论						

2.8.6 注意事项

(1) 薄层板表面应平整、无麻点、气泡或破损、污染等。

(2) 点样时不能损伤薄层板表面。

(3) 注意点样线必须在展开剂上面,不能浸入展开剂中。

2.8.7 思考题

(1) 影响吸附薄层色谱 R_f 值的因素在哪些?解释磺胺甲噁唑与甲氧苄啶的色谱分离原理。

(2) 薄层色谱的常用检测方法有哪些?

2.8 Identification of Compound Sulfamethoxazole Tablets by Thin Layer Chromatography

2.8.1 Objectives

(1) Master the calculation and identification method of the relative shift value of the separated components on the thin layer plate.

(2) Be familiar with the basic operation of thin layer chromatography.

(3) Understand the application of thin layer chromatography in drug identification.

2.8.2 Principles

Sulfamethoxazole and trimethoprim can be separated by using stationary phase adsorption, mobile phase desorption, and the difference in the properties of each component of the mixture. The components can be detected on the thin layer plate due to the absorption of ultraviolet light by the component spots.

2.8.3 Instruments and Reagents

1. Instruments

UV analyzer, silica gel (GF_{254}), chromatography cylinder, mortar, graduated cylinder (25 mL, 500 mL), conical flask (10 mL), funnel (9 cm), filter paper (9 cm), spotting capillary.

2. Reagents

Chloroform, methanol, dimethylformamide, sulfamethoxazole reference standard, trimethoprim reference standard, and compound sulfamethoxazole tablets (the content of sulfamethoxazole in each tablet should be 0.360～0.440 g, and the content of trimethoprim should be 72.0～88.0 mg).

2.8.4 Procedures

1. Preparation of Silica Gel GF_{254} Thin-layer Plate

Weigh 10 g silica gel GF_{254} and place into a mortar, add 20 mL 0.7%

sodium carboxymethyl cellulose solution, grind it in the same direction, and pour it on a glass plate to form a thin layer, put it on a level surface to dry and set aside.

2. Preparation of Reference Solution

(1) Weigh 0.2 g sulfamethoxazole reference standard, add 10 mL methanol to dissolve, and use it as the sulfamethoxazole reference solution.

(2) Weigh 40 mg trimethoprim reference standard, dissolve it in 10 mL methanol, and use it as trimethoprim reference solution.

(3) Weigh 0.2 g sulfamethoxazole reference standard and 40 mg trimethoprim reference standard, dissolve with 10 mL methanol, and use it as a mixed reference substance solution.

3. Preparation of the Test Solution

Weigh an appropriate amount of compound sulfamethoxazole fine powder (approximately equivalent to 0.2 g sulfamethoxazole), add 10 mL methanol to dissolve, shake, filter, and take the filtrate as the test solution.

4. Experiment of Thin Layer Chromatography

Take 5 μL of each of the above solutions, place them on the same silica gel GF_{254} plate, use chloroform : methanol : dimethylformamide (20 : 2 : 1) as the developing solvent, put them in the air to dry, and view them under UV light (254 nm). The position and color of the main spots of the two components displayed by the test solution should be the same as that of the reference solution.

2.8.5 Data Recording and Processing

Record the experimental date in Table 2.6.

Table 2.6

	Reference solution		Mixed reference solution		Test solution	
	Sulfamethoxazole	Trimethoprim	Sulfamethoxazole	Trimethoprim	Spot A	Spot B
Distance from origin to spot center (mm)						

continued

	Reference solution		Mixed reference solution		Test solution	
	Sulfamethoxazole	Trimethoprim	Sulfamethoxazole	Trimethoprim	Spot A	Spot B
Distance from origin to solution front (mm)						
R_f						
Conclusion						

2.8.6 Cautions

(1) The surface of the thin layer plate should be flat, without pitting, bubbles, damage, or pollution, etc.

(2) Do not damage the surface of the thin layer plate when spotting.

(3) The spotting line must be above the developing agent and cannot be immersed in the developing agent.

2.8.7 Questions

(1) What are the factors that affect the R_f value of thin layer chromatography? Explain the principle of chromatographic separation of sulfamethoxazole and trimethoprim.

(2) What are the common detection methods of thin layer chromatography?

2.9 纸色谱法分离氨基酸

2.9.1 实验目的

(1) 掌握纸色谱法分离鉴定氨基酸的原理。
(2) 熟悉纸色谱法的基本操作。
(3) 了解纸色谱法在混合物分离鉴定中的应用。

2.9.2 基本原理

纸色谱法的分离机理主要是分配色谱,固定相是滤纸纤维(载体)上吸附的水,流动相是与水不相混溶的有机溶剂,由于各种氨基酸在结构上存在差异,因此,它们在水中和有机溶剂中溶解度各不相同。极性大的氨基酸在水中溶解度较大,在有机溶剂中溶解度较小,因而有较大的分配系数(K)和较小的比移值(R_f);极性小的氨基酸则正好相反,有较小的分配系数(K)和较大的比移值(R_f),因此达到分离。通过茚三酮显色,便可确定各氨基酸的 R_f 值,并与对照品比较进行定性鉴别。

2.9.3 仪器与试剂

1. 仪器

层析缸(用 $\Phi = 15$ cm 的培养皿代替),新华色谱滤纸(中速),毛细管,电吹风,喷雾器,铅笔,直尺。

2. 试剂

乙氨酸,丙氨酸,蛋氨酸三种对照品溶液(1 mg/mL)。

样品溶液:三种氨基酸混合液。

展开剂:正丁醇∶冰醋酸∶水(4∶1∶2)。

显色剂:0.2%茚三酮正丁醇溶液。

2.9.4 实验步骤

1. 点样

取平整、无折痕的色谱滤纸一张,在中心用铅笔画一直径为 1.5 cm 的圆,对其四等分,作好点样标记。用毛细管分别点上样品溶液和标准溶液,各点 2～3 次。

2. 展开

将滤纸平铺在盛有展开剂(15～20 mL)的层析缸上，预饱和 20 min，穿好纸芯，垂直浸入展开剂中，加盖，自然展开，展开距离约为 5 cm 时，取出，用铅笔画好前沿线，晾干。

3. 显色

喷洒显色剂，电吹风吹干后加热至出现紫色斑点，测量并计算各斑点的 R_f 值。

2.9.5　数据记录与处理

1. 数据记录

将实验数据记入表 2.7。

表 2.7

	对照品溶液			样品溶液		
	乙氨酸	丙氨酸	蛋氨酸	A	B	C
原点至斑点中心的距离(mm)						
原点至溶剂前沿的距离(mm)						
R_f						

2. 实验结果

斑点 A、B、C 分别代表：________________________________。

2.9.6　注意事项

(1) 点样时，必须挥干溶剂才可点第二次，否则点样斑点较大，不利于分离。
(2) 茚三酮能对汗液(含氨基酸)显色，在拿取滤纸时，应保持滤纸清洁。
(3) 喷洒显色剂的量不要过多，避免显色剂溶液在滤纸上流淌。
(4) 标记线不能用圆珠笔和钢笔画，只能用铅笔画。

2.9.7 思考题

(1) 本实验的三种氨基酸 R_f值的大小顺序是什么？为什么这样排？

(2) 为什么纸色谱法用的展开剂多数含有水？

(3) 影响纸色谱 R_f值的因素有哪些？操作中应注意哪些问题？

2.9 Separation of Amino Acids by Paper Chromatography

2.9.1 Objectives

(1) Master the principle of separation and identification of amino acids by paper chromatography.

(2) Be familiar with the operation method of paper chromatography.

(3) Understand the application of paper chromatography in the separation and identification of mixtures.

2.9.2 Principles

The separation mechanism of paper chromatography is mainly partition chromatography, the stationary phase is the water adsorbed on the filter paper fiber (carrier), the mobile phase is an organic solvent that is miscible with water. Due to the structural differences in various amino acids, they have different solubility in water and organic solvents. Amino acids with high polarity have a greater solubility in water and lower solubility in organic solvents, so they have a larger partition coefficient (K) and a smaller specific shift value (R_f). Amino acids with lower polarity are just the opposite, with a smaller partition coefficient (K) and a larger specific shift value (R_f), thus amino acids can be separated. The R_f value of each amino acid can be determined by the color development of ninhydrin and compared with the reference substance for qualitative identification.

2.9.3 Instruments and Reagents

1. Instruments

Chromatographic chamber (replaced with $\Phi = 15$ cm Petri dish), Xinhua chromatography filter paper (medium speed), capillary, hair dryer, sprayer, pencil, ruler.

2. Reagents

Standard/reference solution (1 mg/mL): glycine, alanine, methionine.

Sample solution: three amino acid mixtures.

Developing solvents: n-butanol : glacial acetic acid : water (4 : 1 : 2).
Color developer: 0.2% ninhydrin n-butanol solution.

2.9.4 Procedures

1. Spotting

Take a piece of flat, non-creased chromatographic filter paper, draw a circle with a diameter of 1.5 cm in the center with a pencil, divide it into four equal quarters, and make a spotting mark. Use capillaries to deliver the sample solution and the standard solution respectively, 2 to 3 times each.

2. Developing

Lay the filter paper flat on the chromatographic chamber/developing chamber containing the developing solvent (15 ~ 20 mL). Presaturate for 20 min, impale with the paper antisiphon rod, immerse it in the developing solvent vertically, and then seal the chamber and develop naturally. When the developing distance reaches about 5 cm, open the chamber, remove the paper. Mark the location of the solvent front quickly, and let it dry.

3. Visualization of the Spots

Spray the color developer, dry it with a hair dryer and heat until purple spots appear. Finally, measure and calculate the R_f value of each spot.

2.9.5 Data Recording and Processing

1. Data Recording

Record the experimental data in Table 2.7.

Table 2.7

	Standard solution			Sample solution		
	Glycine	Alanine	Methionine	A	B	C
The distance from the origin to the center of the spot (mm)						
The distance from the origin to the solvent front (mm)						
R_f						

2. Experimental Results

Spots A, B and C represent：____________________________,respectively.

2.9.6 Cautions

(1) When spotting, the test solution must be applied in separate portions to the same spot, and each portion should be allowed to dry before the next is added, otherwise the test spot will be large, which is not conducive for separation.

(2) Ninhydrin can show color on sweat (containing amino acids). When handling the filter paper, keep the filter paper clean.

(3) Do not spray too much color developer to avoid the color developer solution flowing on the filter paper.

(4) The marking line cannot be drawn with ballpoint pens and pens, but only with pencils.

2.9.7 Questions

(1) What is the order of the R_f values of the three amino acids in this experiment? Why is it sorted in this way?

(2) Why do most of the developing solvents used in paper chromatography contain water?

(3) What are the factors that affect the R_f value of paper chromatography? What issues should be noted during operation?

2.10 气相色谱仪性能考察

2.10.1 实验目的

(1) 熟悉气相色谱仪的结构。

(2) 熟悉气相色谱仪的使用方法。

(3) 掌握气相色谱法系统适用性试验的方法。

2.10.2 基本原理

1. 气相色谱仪

气相色谱法是以气体为流动相的色谱法,所用的仪器为气相色谱仪,由载气源、进样部分、色谱柱、检测器和数据处理系统等组成。分析物被气化后,被气体流动相(载气)带入色谱柱进行分离,各组分先后进入检测器,由数据处理系统记录色谱信号。进样部分、色谱柱和检测器的温度均应根据分析要求适当设定。

2. 系统适用性试验

系统适用性试验用于确认与分析方法相关的量测系统和分析操作是否足以适用于预期分析。色谱系统的适用性试验通常包括理论板数、分离度、灵敏度、拖尾因子和重复性,即用规定的对照品溶液或系统适用性试验溶液在规定的色谱系统进行试验,必要时,可对色谱系统进行适当调整,以符合要求。

(1) 色谱柱的理论板数(n)用于评价色谱柱的效能,按下式:

$$n = 16\left(\frac{t_R}{W}\right)^2$$

或

$$n = 5.54\left(\frac{t_R}{W_{h/2}}\right)^2$$

计算色谱柱的理论板数,式中 t_R为色谱峰的保留时间,W 为色谱峰的峰宽,$W_{h/2}$为色谱峰的半峰宽。

(2) 分离度(R)用于评价待测物质与被分离物质之间的分离程度,是衡量色谱系统分离效能的关键指标。待测物质色谱峰与相邻色谱峰之间的分离度应不小于1.5。分离度的计算公式为

$$R = \frac{2 \times (t_{R_2} - t_{R_1})}{W_1 + W_2}$$

式中，t_{R_2}为相邻两色谱峰中后一峰的保留时间，t_{R_1}为相邻两色谱峰中前一峰的保留时间，W_1、W_2分别为此相邻两色谱峰的峰宽。

(3) 灵敏度用于评价色谱系统检测微量物质的能力，通常以信噪比(S/N)来表示。

(4) 拖尾因子(T)用于评价色谱峰的对称性。拖尾因子计算公式为

$$T = \frac{W_{0.05h}}{2d_1}$$

式中，$W_{0.05h}$为5%峰高处的峰宽，d_1为峰顶在5%峰高处横坐标平行线的投影点至峰前沿与此平行线交点的距离，T值应在0.95～1.05之间。

(5) 重复性用于评价色谱系统连续进样时响应值的重复性能。通常取对照品溶液连续进样5次，其峰面积测量值的相对标准偏差应不大于2.0%。

2.10.3　仪器与试剂

1. 仪器

气相色谱仪(配备自动进样器、氢火焰离子化检测器)，分析天平，微量注射器。

2. 试剂

苯，甲苯，丙酮(均为分析纯试剂)。

2.10.4　实验步骤

1. 溶液制备

以丙酮为溶剂，分别配制100 mg/L的苯标准溶液、100 mg/L的甲苯标准溶液、100 mg/L的苯和甲苯混合标准溶液。

2. 色谱条件设置

选择固定相为5%苯基、95%二甲基聚硅氧烷的毛细管柱(30 m×0.25 mm×0.25 μm)或其他适合的弱极性色谱柱，载气为氮气，柱内流量为1 mL/min，分流比为1∶50，进样口温度为200 ℃，检测器温度为220 ℃，柱温为80 ℃，氢气流速为30 mL/min，空气流速300 mL/min(色谱条件可根据实际仪器情况进行优化调节)。

3. 系统适用性试验

在上述色谱条件下，分别吸取苯标准溶液、甲苯标准溶液各1 μL进行气相色谱仪进样，记录其各自色谱峰的保留时间；吸取1 μL混合标准溶液进样，记录各色谱峰的保留时间、峰面积，计算理论板数、分离度，并连续进样5次考察重复性。

2.10.5 数据记录与处理

1. 数据记录

将实验数据记入表 2.8。

表 2.8

	苯标准溶液	甲苯标准溶液	混合标液 1	混合标液 2	混合标液 3	混合标液 4	混合标液 5
t_{R_1}		—					
t_{R_2}	—						
W_1		—					
W_2	—						

2. 结果计算

按以下公式计算结果:

$$n = 16\left(\frac{t_R}{W}\right)^2$$

$$R = \frac{2 \times (t_{R_2} - t_{R_1})}{W_1 + W_2}$$

$$RSD(\%) = \frac{W_{mean}}{SD} \times 100$$

2.10.6 注意事项

(1) 开启气相色谱仪电源前,必须先通载气;基线达到平衡后,方可进样分析;试验结束后,先将柱温降至所要求温度后才可以关闭电源,最后关闭载气。

(2) 大多数色谱柱在 100 ℃以上的温度下使用时,都不能很好地耐受水和氧气。必须特别注意载气纯度以及气源中的杂质,载气需使用高纯度氮气。

(3) 新的色谱柱应进行老化,以除去柱内残留的溶剂、键合或者涂布不够稳定的固定相,让色谱柱性能更稳定。

(4) 进样口的硅橡胶垫经过几十次进样后,容易漏气,需及时更换。

2.10.7 思考题

(1) 自动进样相对于手动进样具有什么优势?

(2) 为什么采用毛细管柱时需要进行分流进样?

2.10 Performance Examination of the Gas Chromatograph System

2.10.1 Objectives

(1) Get acquainted with the structure of the gas chromatograph system.

(2) Be familiar with the usage of the gas chromatograph system.

(3) Master the method of the chromatograph system suitability test.

2.10.2 Principles

1. Instrumentation of Gas Chromatograph

Gas chromatography is a method which uses gas as the mobile phase. A gas chromatographic system consists of a carrier gas, an injection, a column, a detector, and a data system. After the analytes are vaporized, they are carried into the column by the gas mobile phase (carrier gas) for separation and the chromatographic signal is recorded by the data processing system when each component enters the detector successively. The temperature of the injection, column and detector should all be set appropriately according to the analytical requirements.

2. System Suitability Test (SST)

The system suitability test is used to verify that the measurement system and analytical operation associated with the analytical procedure are adequate for the intended analysis. The system suitability test of the chromatograph usually includes theoretical plate number, resolution, sensitivity, tailing factor and reproducibility. SST is performed in the specified chromatographic system with the specified control solution or system suitability test solution, and if necessary, the chromatographic system can be adjusted appropriately to meet the requirements.

(1) The theoretical plate number (n) of the column was used to evaluate the performance of the column. The theoretical plate number of the column was calculated according to

$$n = 16\left(\frac{t_R}{W}\right)^2 \quad \text{or} \quad n = 5.54\left(\frac{t_R}{W_{h/2}}\right)^2$$

Where t_R is the retention time of the chromatographic peak, W is the peak width of the chromatographic peak, and $W_{h/2}$ is the half-peak width of the chromatographic peak.

(2) Degree of separation (R) is used to evaluate the degree of separation between the substance to be measured and the substance to be separated, and is a key index to measure the separation efficiency of the chromatographic system. The separation degree between chromatographic peaks of the substance to be measured and adjacent chromatographic peaks should not be less than 1.5. The separation degree can be calculated as follows:

$$R = \frac{2 \times (t_{R_2} - t_{R_1})}{W_1 + W_2}$$

Where t_{R_2} is the retention time of the last peak of the two adjacent chromatographic peaks; t_{R_1} is the retention time of the previous peak of two adjacent chromatographic peaks; W_1 and W_2 were the peak widths of two adjacent chromatographic peaks, respectively.

(3) Sensitivity is used to evaluate the ability of chromatographic systems to detect trace substances, which is usually expressed as the signal-to-noise ratio (S/N).

(4) The trailing factor (T) was used to evaluate the symmetry of chromatographic peaks. The formula for calculating the trailing factor is

$$T = \frac{W_{0.05h}}{2d_1}$$

Where $W_{0.05h}$ is the peak width of 5% peak height; d_1 is the distance from the projection point of the horizontal parallel line at the peak height of 5% to the intersection point of the peak front and the parallel line, and the T value should be between 0.95 and 1.05.

(5) Repeatability is used to evaluate the repeatability of response value during continuous injection of chromatographic system. The reference solution is usually taken for 5 consecutive injections, and the relative standard deviation of the measured peak area should be no more than 2.0%.

2.10.3 Instruments and Reagents

1. Instruments

Gas chromatograph (with autosampler, hydrogen flame ionization detector), analytical balance, microinjector.

2. Reagents

Benzene, toluene, acetone (all are analytical pure reagents).

2.10.4 Procedures

1. Solution Preparation

Using acetone as the solvent, prepare 100 mg/L benzene standard solution, 100 mg/L toluene standard solution and 100 mg/L benzene and 100 mg/L toluene mixed standard solutions, respectively.

2. Setting of Gas Chromatographic Conditions

Use a capillary column (30 m×0.25 mm×0.25 μm) with 5% phenyl and 95% dimethylpolysiloxane as stationary phase or another suitable weakly polar column. The carrier gas is nitrogen and the flow rate in the column is 1 mL/min. The splitting ratio is 1 : 50. The inlet temperature, the column temperature and the detector temperature are 200 ℃, 80 ℃ and 220 ℃, respectively. The hydrogen flow rate is 30 mL/min, and the air flow rate is 300 mL/min. (The chromatographic conditions can be optimized according to the actual instrument conditions.)

3. System Suitability Test

Under the above set chromatographic conditions, inject 1 μL of the benzene standard solution and toluene standard solution into the gas chromatograph system respectively, and record the retention time of their respective peaks. Inject 1 μL of mixed standard solution into the gas chromatograph system, and record the retention time and peak area of each peak to calculate the theoretical plate number, separation, and trailing factor. The reproducibility is investigated by repeat injection five times continuously.

2.10.5 Data Recording and Processing

1. Data Recording

Record the experimental data in Table 2.8.

Table 2.8

	Benzene standard solution	Toluene standard solution	Mixed standard solution 1	Mixed standard solution 2	Mixed standard solution 3	Mixed standard solution 4	Mixed standard solution 5
t_{R_1}		—					
t_{R_2}	—						
W_1		—					
W_2	—						

2. Results Calculation

Calculate the results according to the following formula:

$$n = 16\left(\frac{t_R}{W}\right)^2$$

$$R = \frac{2 \times (t_{R_2} - t_{R_1})}{W_1 + W_2}$$

$$RSD(\%) = \frac{W_{mean}}{SD} \times 100$$

2.10.6 Cautions

(1) The carrier gas must be passed before turning on the power of the gas chromatograph system. The sample can be analyzed while the baseline reaches equilibrium. When the test is finished, the column temperature should be lowered to the required temperature before turning off the power, and finally the carrier gas must be turned off.

(2) Most columns do not tolerate moisture and oxygen well when operated at temperatures over 100 ℃. Special attention must be given to the purity of the source and the gas chromatograph and to impurities in the gas supply. High purity nitrogen is required for carrier gas.

(3) New columns should be aged to remove residual solvents, bonded or coated stationary phases that are not stable enough in the column to make the column performance more stable.

(4) The silicone rubber pad of the sample inlet is prone to air leakage after dozens of injections and needs to be replaced in time.

2.10.7 Questions

(1) What are the advantages of auto injection over manual injection?

(2) Why is the split injection necessary when using a capillary column?

2.11 高效液相色谱仪基本操作与系统适应性

2.11.1 实验目的

(1) 熟悉高效液相色谱仪的构造及基本操作。
(2) 掌握色谱柱理论塔板数、分离度、拖尾因子和重复性的计算方法。
(3) 了解考察色谱柱基本特性的方法和指标。

2.11.2 基本原理

色谱系统的系统适用性试验通常包括理论塔板数、分离度、拖尾因子和重复性四个参数。

1. 理论塔板数(N)

根据塔板理论,理论塔板数越大,板高越小,色谱柱的柱效能越高,所以理论塔板数一般用于评价色谱柱的分离效能。计算公式如下:

$$N = \frac{L}{H} = \frac{16t_R^2}{W^2}$$

式中,L 为色谱柱柱长,H 为理论塔板高度,t_R 为组分的保留时间,W 为组分的峰宽。

2. 分离度(R)

从色谱峰判断相邻两组分在色谱柱中被分离状况的指标,计算公式如下:

$$R = \frac{2(t_{R_2} - t_{R_1})}{W_1 + W_2}$$

式中,t_{R_2} 为相邻两峰后一峰的保留时间,t_{R_1} 为相邻两峰前一峰的保留时间,W_1 和 W_2 为相邻两峰的峰宽。

3. 拖尾因子(T)

色谱柱的热力学性质和柱填充均匀与否的指标,用于评价色谱峰的对称性,当 T 位于 0.95~1.05 范围时,为对称色谱峰。计算公式如下:

$$T = \frac{W_{0.05h}}{2A}$$

式中,$W_{0.05h}$ 为组分位于 0.05 倍峰高处的峰宽,A 为色谱峰前沿与色谱峰顶点至基线的垂线之间的距离。

4. 重复性

用于评价色谱系统在连续进样时响应值的重复性能。

考察各类型色谱柱性能的常用化合物和操作条件见表2.9。

表2.9　色谱柱类型及对应的操作条件

色谱柱类型	测试用化合物	流动相
吸附柱	苯、萘、联苯	己烷或庚烷
反相柱	苯、萘、菲、联苯	甲醇：水(80：20)
氰基柱	甲苯、苯乙腈、二苯酮	己烷：异丙醇(98：2)
氨基柱	联苯、菲、硝基苯	庚烷或异辛烷
醚基柱	邻、间、对-硝基苯胺	己烷：二氯甲烷：异丙醇(70：30：5)

2.11.3　仪器与试剂

1. 仪器

高效液相色谱仪(配紫外检测器)，微量注射器(100 μL)，超声波清洗器，滤膜(0.45 μm)及抽滤装置，色谱柱：C_{18}反相键合色谱柱(250 mm×4.6 mm)。

2. 试剂

苯、萘、菲、联苯、甲醇(均是色谱纯)，重蒸馏水(新制)。

2.11.4　实验步骤

1. 仪器使用方法

色谱条件：

流动相为甲醇：水(80：20)，流速：1 mL/min，固定相：C_{18}反相键合色谱柱，检测波长：254 nm，进样量：20 μL(定量环)。

2. 配制流动相

按甲醇和水80：20的体积比，量取适量溶剂混合，然后过滤脱气。

3. 准备供试品

配制浓度均为1 μg/mL的苯、萘、菲、联苯的甲醇溶液，作为供试品。

4. 进样测定

用微量注射器吸取四个样品，待基线平直稳定后，注入色谱仪，记录色谱图。选取苯供试品溶液，连续进样5次，记录峰面积 A。

5. 实验复原

实验完毕，根据要求，用甲醇冲洗色谱柱后，关机。

2.11.5 数据记录与处理

根据色谱图信息将相应数据记录到表 2.10 中,经计算得到理论塔板数 N、分离度 R、拖尾因子 T 及重复性结果。

表 2.10

参数＼样品	苯	萘	菲	联苯
保留时间 t_R(min)				
峰宽 W				
峰面积 A				
峰高 h				
$W_{0.05h}$				
理论塔板数 N				
分离度 R				
拖尾因子 T				

重复性的数据及结果

进样次数	1	2	3	4	5
苯的峰面积 A					
RSD(%)					

2.11.6 注意事项

(1) 在色谱仪运行过程中,注意防止流动相流完,废液瓶要及时处理,以免废液溢出。

(2) 注射器使用前要先用待测溶液洗涤数次,进样前排除注射器中气泡,进样时迅速扳动送样手柄。

2.11.7 思考题

(1) 说明苯、萘、菲、联苯在反相色谱中的洗脱顺序及原因。

(2) 流动相在使用前为何要进行脱气处理?

2.11 Basic Operation and System Adaptability of High-performance Liquid Chromatography

2.11.1 Objectives

(1) Be familiar with the structure and basic operation of high-performance liquid chromatography.

(2) Master the calculation method of theoretical plate number, resolution, tailing factor and repeatability of chromatographic column.

(3) Learn about the methods and indexes for examining the basic properties of a chromatographic column.

2.11.2 Principles

The system suitability test of a chromatographic system usually includes four parameters: the number of theoretical plates, resolution, tailing factor and repeatability.

1. Number of Theoretical Plates (*N*)

According to the plate theory, the larger the number of theoretical plates, the smaller the plate height, and the higher the column efficiency of the chromatographic column. The theoretical plate number is generally used to evaluate the separation efficiency of a chromatographic column. Calculated as follows:

$$N = \frac{L}{H} = \frac{16t_R^2}{W^2}$$

Where L is the length of a column, H is the plate height, t_R is the retention time of the peak and W is the peak width of the component.

2. Resolution (*R*)

An index for determining the separation status of two adjacent components in a chromatographic column from chromatographic peaks. Calculated as follows:

$$R = \frac{2(t_{R_2} - t_{R_1})}{W_1 + W_2}$$

Where t_{R_2} is the retention time of the last of the two adjacent peaks, t_{R_1} is the retention time of the previous peak of the two adjacent peaks; W_1 and W_2 are the peak widths of two adjacent peaks.

3. Tailing Factor (T)

Thermodynamic properties of a chromatographic column and indicators of uniformity of column packing. Tailing factor is used to evaluate the symmetry of chromatographic peaks. When T is in the range of 0.95~1.05, chromatographic peaks are symmetrical peaks. Calculated as follows:

$$T = \frac{W_{0.05h}}{2A}$$

Where $W_{0.05h}$ is the peak width of the component at 0.05 times the peak height, A is the distance between the front of the chromatographic peak and the vertical line from the apex of the chromatographic peak to the baseline.

4. Repeatability

Used to evaluate the repeatability of the response value of a chromatographic system during continuous injection.

Common compounds and operating conditions for examining the performance of each type of chromatographic column are shown in Table 2.9.

Table 2.9 Column types and corresponding operating conditions

Column type	Test compound	Mobile phase
Adsorption column	Benzene, Naphthalene, Biphenyl	Hexane or Heptane
Reversed-phase column	Benzene, Naphthalene, Phenanthrene, Biphenyl	Methanol : Water(80 : 20)
Cyano column	Toluene, Phenylacetonitrile, Benzophenone	Hexane : Isopropanol(98 : 2)
Amino column	Biphenyl, Phenanthrene, Nitrobenzene	Heptane or Isooctane
Ether-based column	o-, m-, p-Nitroaniline	Hexane : Dichloromethane : Isopropanol(70 : 30 : 5)

2.11.3 Instruments and Reagents

1. Instruments

High-performance liquid chromatograph (equipped with UV detector), microsyringe (100 μL), ultrasonic cleaner, filter membrane (0.45 μm) and suction filter device, chromatography column: C_{18} bonded reversed-phase column (250 mm×4.6 mm).

2. Reagents

Benzene, naphthalene, phenanthrene, biphenyl and methanol (chromatographically pure), double distilled water (newly prepared).

2.11.4 Procedures

1. Operation Method of HPLC

Conditions:

Mobile phase: methanol : water (80 : 20), velocity of flow: 1 mL/min, stationary phase: C_{18} bonded reversed-phase column, detection wavelength: 254 nm, injection volume: 20 μL (quantitative loop).

2. Preparation of Mobile Phase

The volume ratio of methanol and water is 80 : 20. Measure appropriate amount of solvent to mix, then filter and degas.

3. Preparation of the Test Article

Preparing the following solution as the test subject: methanol solution of benzene, naphthalene, phenanthrene and biphenyl. with the concentration of 1 μg/mL.

4. Measurement

Draw the four samples with a microsyringe, and after the baseline is flat and stable, inject it into the chromatograph and record the chromatogram. Select the benzene test solution, inject 5 times continuously, and record the peak area A.

5. Experimental Recovery

After the experiment is completed, according to the requirements, flush the column with methanol, then shut down the instruments.

2.11.5 Data Recording and Processing

According to the chromatogram information, the corresponding data is recorded in Table 2.10. Calculate the theoretical plate number N, resolution R, tailing factor T, and repeatability results are obtained by calculation.

Table 2.10

Samples / Parameters	Benzene	Naphthalene	Phenanthrene	Biphenyl
Retention time t_R(min)				
Peak width (W)				
Peak area (A)				
Peak height (h)				
$W_{0.05h}$				
Theoretical plate number (N)				
Resolution (R)				
Tailing factor (T)				

Repeatability data and results

Number of injections	1	2	3	4	5
Peak area of benzene (A)					
RSD(%)					

2.11.6 Cautions

(1) During the operation of the chromatography, pay attention to preventing the mobile phase from running out, and dispose of the waste liquid bottle in time to prevent the waste liquid from overflowing.

(2) The syringe should be washed several times with the solution to be sucked. The air bubbles in the syringe should be removed before injection, and the sample delivery handle should be moved quickly during injection.

2.11.7 Questions

(1) Explain the elution order of benzene, naphthalene, phenanthrene and biphenyl in reversed-phase chromatography and give reasons.

(2) Why should the mobile phase be degassed before use?

2.12 高效液相色谱法定性分析

2.12.1 实验目的

(1) 掌握高效液相色谱仪的仪器构造。
(2) 掌握高效液相色谱仪的定性方法。

2.12.2 基本原理

每一种化合物在特定的色谱条件下(流动相组成、配比、色谱柱、柱温等),具有特定的保留时间,因此可以利用保留时间进行定性分析。如果在相同的色谱条件下,被测化合物与对照溶液的保留时间一致,就可以初步认为被测化合物与标准样品相同,从而实现对未知化合物的定性分析。

2.12.3 仪器与试剂

1. 仪器

高效液相色谱仪(紫外检测器),C_{18}反相键合色谱柱(150 mm×4 mm),微量进样器(25 μL),过滤器(0.45 μm)及脱气装置。

2. 试剂

苯,甲苯,萘,甲醇(色谱纯),高纯水(新制)。

2.12.4 实验步骤

1. 色谱条件

流动相为甲醇∶水(80∶20),固定相为C_{18}反相键合色谱柱,检测波长为254 nm,流量1 mL/min。

2. 供试品溶液配制

配制苯、甲苯、萘的混合甲醇溶液(1 μg/mL),作为供试品溶液。

3. 对照品溶液制备

分别配制苯、甲苯、萘的甲醇溶液(1 μg/mL),作为对照溶液。

4. 流动相

甲醇∶水(80∶20)混合后过滤,并进行超声脱气。

5. 测定

吸取供试品溶液 25 μL，赶气泡后注入色谱仪，记录色谱图。在相同条件下，分别吸取对照品苯、甲苯、萘溶液 25 μL 依次进样，记录色谱图。将供试液色谱图中各峰的保留时间与各对照品的保留时间比较，找出各峰的归属。

2.12.5　数据记录与处理

1. 数据记录

将实验数据记入表 2.11。

表 2.11

	对照品溶液			样品溶液		
组分名称	苯	甲苯	萘	峰 1	峰 2	峰 3
保留时间 t_R(min)						

2. 实验结果

样品溶液中峰 1、2、3 分别对应的化合物是：________、________、________。

2.12.6　思考题

(1) 利用高效液相色谱仪定性的方法有几种？分别在什么情况下适用？
(2) 影响组分保留值的因素有哪些？
(3) 试说明苯和甲苯出峰顺序的理论依据。

2.12 Qualitative Analysis by High Performance Liquid Chromatography

2.12.1 Objectives

(1) Master the instrument structure of high performance liquid chromatography (HPLC).

(2) Master the qualitative method of HPLC.

2.12.2 Principles

Each compound has a specific retention time under particular chromatographic conditions (such as composition and proportion of mobile phase, chromatographic column, column temperature, etc.). Therefore, the retention time can be used for qualitative analysis of compounds. If the retention time of the tested compound is consistent with that of the control solution under the same chromatographic conditions, it can be preliminarily considered that the tested compound is the same as the standard sample. Thus, the qualitative analysis of unknown compounds can be realized.

2.12.3 Instruments and Reagents

1. Instruments

High performance liquid chromatography (with UV detector), C_{18} bonded reversed-phase column (150 mm×4 mm), microsyringe (25 μL), filter (0.45 μm) and degassing device.

2. Reagents

Benzene, toluene, naphthalene, methanol (chromatography pure), ultra pure water (fresh preparation).

2.12.4 Procedures

1. Chromatographic Conditions

The mobile phase should is methanol : water (80 : 20), the stationary phase is C_{18} reverse-phase column. The detection wavelength is 254 nm, and flow rate is 1 mL/min.

2. Preparation of Test Solution

Prepare the mixed methanol solution of benzene, toluene, and naphthalene (1 μg/mL) as test solution.

3. Preparation of Control Solution as Test Solution

Prepare benzene, toluene, and naphthalene solution (1 μg/mL) respectively as control solutions.

4. Mobile Phase

Mix ethanol : water (80 : 20), filter and ultrasonic degas for standby application.

5. Determination

After eliminating air bubbles, 25 μL test solution should be injected into the chromatograph for analysis. Under the same conditions, 25 μL of benzene, toluene and naphthalene solutions are absorbed and injected respectively, and the chromatograms are recorded. The retention time of each peak in the chromatogram of the test solution should be compared with that of each reference substance to find out the attributes of each peak. The compounds corresponding to each chromatographic peak can be found according to the retention time.

2.12.5 Data Recording and Processing

1. Data Recording

Record the experimental date in Table 2.11.

Table 2.11

	Reference solution			Test solution		
Compound	Benzene	Methyl-benzene	Naphth-alene	Peak 1	Peak 2	Peak 3
Retention time t_R(min)						

2. Experimental Results

Peaks 1, 2 and 3 in the sample solution correspond to: ________, ________, ________。

2.12.6 Questions

(1) How many qualitative methods can be used by HPLC? Under what circumstances are they applicable?

(2) What are the factors affecting the retention value?

(3) Try to explain the theoretical basis of the peak sequence of benzene and toluene.

2.13　毛细管区带电泳分离手性药物的对映异构体

2.13.1　实验目的

(1) 掌握毛细管区带电泳的基本原理。
(2) 了解毛细管电泳仪的基本构造与操作步骤。
(3) 了解毛细管电泳法在手性药物分离中的应用。

2.13.2　基本原理

毛细管区带电泳(capillary zone electrophoresis,CZE)是毛细管电泳中最基本也是应用较广的一种分离方法。它主要是根据物质电泳迁移率的差异实现物质的分离。物质的迁移速度可以用下列公式表示:

$$u_{ap} = (u_{ep} + u_{eo})E$$

式中,u_{ap}为物质的表观迁移速度,u_{ep}为物质的电泳速度,u_{eo}为物质的电渗速度。

当毛细管内壁带负电时,带正电荷物质的电泳和电渗流移动方向相同,带负电荷物质的电泳和电渗流移动方向相反,对于中性物质只有电渗流速度,所以洗脱顺序是:阳离子物质、中性物质、阴离子物质。当毛细管内壁带正电时,洗脱顺序刚好相反。但是这两种方式都无法分离中性分子。

毛细管区带电泳法分离手性药物对映体的原理为:在电解质缓冲液中添加手性选择剂,由于两个对映体和手性选择剂构成三点作用模式形成的包容络合物的稳定性不同,即络合常数不同,使得它们的表观迁移率产生差异,因而使对映体得到分离。环糊精类化合物是一类常用的手性选择剂。

氧氟沙星和左氧氟沙星都是广谱抗菌的氟喹诺酮类药物,两者在杀菌方面的作用机制相同,但是左氧氟沙星对大多数细菌的抗菌活性为氧氟沙星的2倍,且两者的适应证有所区别。左氧氟沙星分子的结构式如图2.1所示。

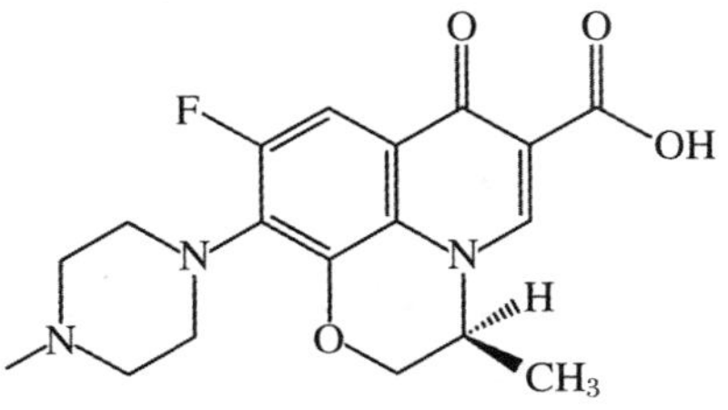

图2.1　左氧氟沙星分子的结构式

2.13.3 仪器与试剂

1. 仪器

毛细管电泳仪,pH 计,超声波清洗机,0.45 μm 滤膜。

2. 试剂

二甲基-β-环糊精,H_3PO_4,KH_2PO_4,NaOH(均为分析纯),二次去离子水,对照品:氧氟沙星和左氧氟沙星。

2.13.4 实验步骤

1. 仪器使用方法

参见相关资料。

2. 溶液的配制

(1) 缓冲液的配制:称取适量 KH_2PO_4 配制成 70 mmol/L 溶液,利用 pH 计,用 H_3PO_4 将溶液 pH 调节到 2.5。

(2) 背景电解质溶液的配制:称取适量二甲基-β-环糊精溶于上述缓冲液中,二甲基-β-环糊精的浓度为 40 mmol/L。溶液用 0.45 μm 微孔滤膜过滤,超声波脱气备用。

(3) 样品溶液的配制。分别精密称取适量氧氟沙星和左旋氧氟沙星,加入二次去离子水溶解,配成浓度为 1.5 mg/mL 的溶液,用 0.45 μm 微孔滤膜过滤,超声波脱气备用。

3. 电泳条件

二甲基-β-环糊精溶液:40 mmol/L,电泳缓冲液:70 mmol/L,混合磷酸盐缓冲液(pH 2.5),分离电压:20 kV,检测波长:380 nm,柱温:25 ℃。

4. 进样分离

(1) 毛细管清洗。毛细管柱依次用 0.1 mmol/L NaOH、二次去离子水、背景电解质缓冲溶液各冲洗 10 min(此时不施加高压电源),加电压平衡 5 min。

(2) 进样分析。采用重力进样,进出口两端高度差为 10 cm,进样时间为 10 s。进样后,迅速移开试样管,再换上缓冲溶液池,施加 20 kV 电压进行分离。

2.13.5 数据记录与处理

(1) 记录氧氟沙星和左氧氟沙星样品的手性分离电泳图谱。

(2) 按照面积归一化法分别计算氧氟沙星样品中左旋体和右旋体所占的比例。

2.13.6　注意事项

（1）对于紫外检测，需制作检测窗口，将毛细管小心穿过光学检测器，对准光路，安装好检测池，并将毛细管两端分别插入缓冲液池。

（2）通电后的毛细管勿用手触碰，保持仪器室的湿度和温度条件。两缓冲液池中溶液液面应保持同一水平面。

（3）将样品溶液装入塑料小管时，应防止内壁产生气泡，可用手指轻弹以排除气泡，以免进样时引入空气。

（4）每次进样前，毛细管柱依次用 NaOH 溶液、水、缓冲液清洗 5 min，待基线平稳后再进样。实验完成后要用水清洗毛细管，以防毛细管堵塞。

2.13.7　思考题

（1）毛细管电泳法在药物分析领域有哪些应用？

（2）目前毛细管电泳技术有哪些优点和局限性？

2.13 Separation of Enantiomers of Chiral Drugs by Capillary Zone Electrophoresis

2.13.1 Objectives

(1) Master the principle of capillary zone electrophoresis.

(2) Understand the structure and operation steps of capillary electrophoresis instrument.

(3) Understand the application of capillary electrophoresis in the separation of chiral drugs.

2.13.2 Principles

Capillary zone electrophoresis (CZE) is the most basic and widely used separation method in capillary electrophoresis. Separation is based on differences in electrophoretic mobility. The speed of migration of a substance can be expressed by the following formula:

$$u_{ap} = (u_{ep} + u_{eo})E$$

Where u_{ap} is the apparent velocity, u_{ep} is the electrophoretic mobility, u_{eo} is the electroosmotic mobility.

If the capillary wall is negative, electrophoretic, and electroosmotic flows of positively charged species move in the same direction. Electrophoretic and electroosmotic flow of negatively charged species move in opposite directions. Neutral substances only have electroosmotic flow velocity. Therefore, the order of elution is: cations before neutrals before anions. If the capillary wall charge is reversed, the order of elution is anions substance, neutrals substance, and cations substance. Neither scheme separates neutral molecules from one another.

The principle of separation of chiral drug enantiomers by capillary zone electrophoresis is as follows: when the chiral selectors are added to electrolyte buffers, since the stability of inclusive complexes formed due to the three-point mode of action of the two enantiomers and the chiral selector is different. That is, the complexation constants are different. Their apparent mobilities are different, which enables the separation of enantiomers. Cyclodextrins are a class of commonly used chiral selectors.

Ofloxacin and levofloxacin are both broad-spectrum antibacterial fluoroquinolones. Both of which have the same mechanism of action in sterilization, but levofloxacin's antibacterial activity against most bacteria is twice that of ofloxacin. And the adaptation of the two symptoms is different. The structural formula of levofloxacin molecule is as follows (Figure 2.1).

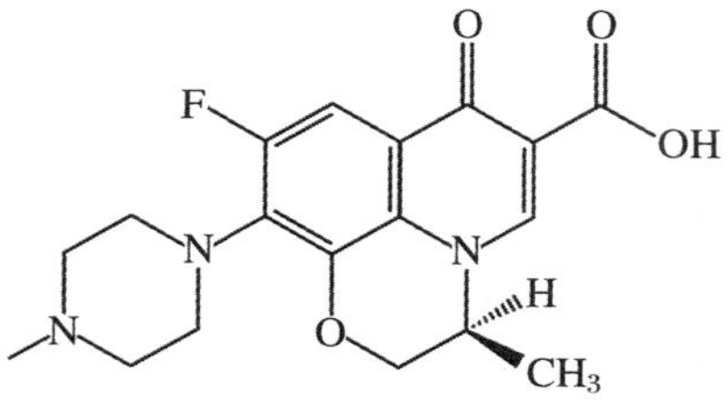

Figure 2.1　Molecular structure of levofloxacin

2.13.3　Instruments and Reagents

1. Instruments

Capillary electrophoresis instrument (CE), pH meter, ultrasonic cleaner, filter membrane (0.45 μm).

2. Reagents

Heptapkis (2,6-di-o-methyl)-β-cyclodextrin, H_3PO_4, KH_2PO_4 and NaOH (analytical pure), double distilled water (fresh), reference substance: ofloxacin and levofloxacin.

2.13.4　Procedures

1. Operation Method of CE

See the relevant information.

2. Prepare the Solution

(1) Preparation of buffer solution. Weigh an appropriate amount of KH_2PO_4 to prepare a solution with a concentration of 70 mmol/L. Using a pH meter, adjust the pH of the solution to 2.5 with H_3PO_4.

(2) Preparation of background electrolyte buffer solution. Weigh an appropriate amount of dimethyl-β-cyclodextrin and add it to the prepared buffer solution. The concentration of dimethyl-β-cyclodextrin is 40 mmol/L. Filter the

solution with a 0.45 μm microporous membrane, ultrasonically degassed for later use.

(3) Preparation of sample solutions. Precisely weigh appropriate amounts of ofloxacin and levofloxacin respectively. Add double distilled water to obtain solution with a concentration of 1.5 mg/mL. Filter with a 0.45 μm microporous membrane, ultrasonically degassed for later use.

3. Electrophoresis Conditions

Dimethyl-β-cyclodextrin solution: 40 mmol/L, running buffer: 70 mmol/L mixed phosphate buffer (pH 2.5), separation voltage: 20 kV, detection wavelength: 380 nm, column temperature: 25 ℃.

4. Sample Separation

(1) Capillary cleaning. The capillary column must be washed sequentially with 0.1 mmol/L NaOH, secondary deionized water, and background electrolyte buffer solution for 10 min each (without voltage). Pressurized equilibration for 5 min.

(2) Sample analysis. Adopt gravity injection. Set the height difference between the inlet and outlet to 10 cm. The injection time is 10 s. After injection, the sample tube must be quickly removed, replaced with a buffer solution cell, and a 20 kV voltage applied for separation.

2.13.5 Data Recording and Processing

(1) Record chiral separation electropherograms of ofloxacin and levofloxacin samples.

(2) According to the area normalization method, the proportions of levorotatory and dextrorotary forms in ofloxacin samples should be calculated respectively.

2.13.6 Cautions

(1) For UV detection, it is necessary to make a detection window, carefully pass the capillary tube through the optical detector, align it with the optical path, install the detection pool, and insert both ends of the capillary tube into the buffer pool.

(2) Do not touch the capillary after turning the power on, and maintain the humidity and temperature conditions of the instrument room. The solution

level in the two buffer pools should be kept at the same level.

(3) When filling the sample solution into the plastic vial, the inner wall should be prevented from generating air bubbles, and the air bubbles can be removed by flicking with fingers to avoid introducing air during injection.

(4) Before each injection, the capillary column must be sequentially washed with NaOH solution, water, and buffer for 5 min, and the injection should be performed after the baseline is stable. After the experiment is completed, the capillary should be washed with water to prevent the capillary from clogging.

2.13.7 Questions

(1) What are the applications of capillary electrophoresis in the field of drug analysis?

(2) What are the advantages and limitations of current capillary electrophoresis techniques?

第 3 章　综合性仪器分析实验

3.1　循环伏安法检测对乙酰氨基酚

3.1.1　实验目的

(1) 学习循环伏安法原理。

(2) 掌握循环伏安法研究对乙酰氨基酚电化学氧化机理过程。

(3) 学习循环伏安法测定对乙酰氨基酚片的含量。

3.1.2　基本原理

伏安法是根据伏安特性曲线进行定性或定量分析的一种电化学分析方法。在电极上施加的可变电位称作激励信号。伏安法中最常见的四种分析方法包括线性扫描伏安法、差分脉冲伏安法、方波伏安法和循环伏安法。其中，电位激励信号若为三角波信号，则获得的伏安曲线就是循环伏安曲线。如图 3.1 所示，扫描电压呈等腰三角形。在电压上升过程中，若化合物在电极表面被氧化，则发生氧化反应；在电压下降过程中，若氧化产物在电极表面发生还原，则发生还原反应。因此，一次三角波扫描经历了氧化和还原过程，故称为循环伏安法。

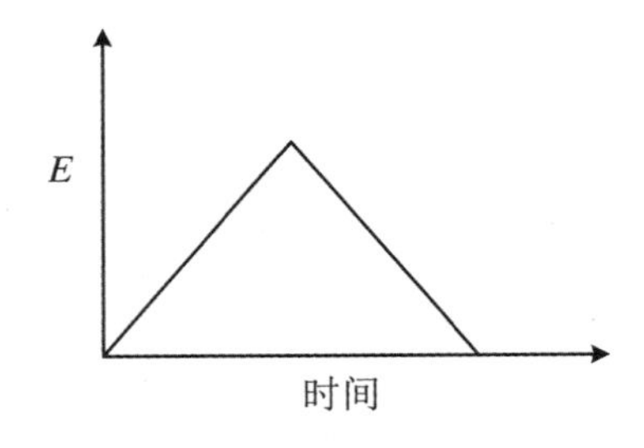

图 3.1　电流-电压曲线

氧化还原峰电位值及其差值和氧化还原峰电流值，是循环伏安法中重要的参数。如图3.2所示，对于一个电极反应，若阳极峰电位（E_{ap}）与阴极峰电位（E_{cp}）的峰电位差值（ΔE_p）接近0.059/n（V，25 ℃），阳极峰电流和阴极峰电流在数值上近似相等，则该反应是可逆反应。当反应不可逆时，ΔE_p差值越大，不可逆程度越大。

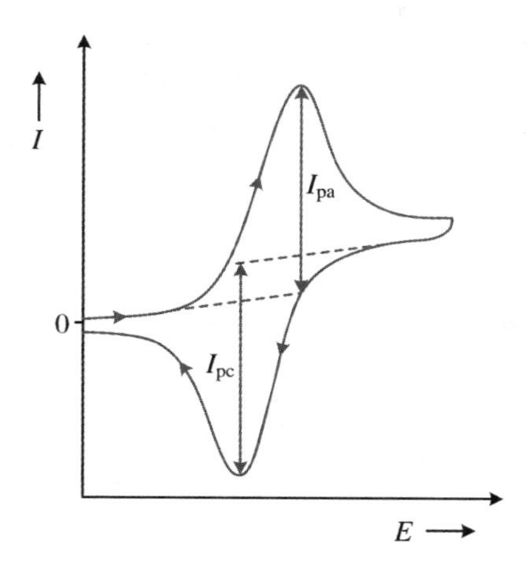

图3.2　电流-电压曲线

对乙酰氨基酚（N-acetyl-para-aminophenol，APAP）是非甾体抗炎药的重要成分，主要是通过抑制前列腺素的合成起到解热镇痛药的作用。循环伏安法可用于研究电活性在电极表面的电子转移过程，是研究反应机理的有效手段。本实验利用循环伏安法研究APAP在电极表面的反应机理及对乙酰氨基酚片中APAP含量。APAP在电极表面的反应机理如图3.3所示。

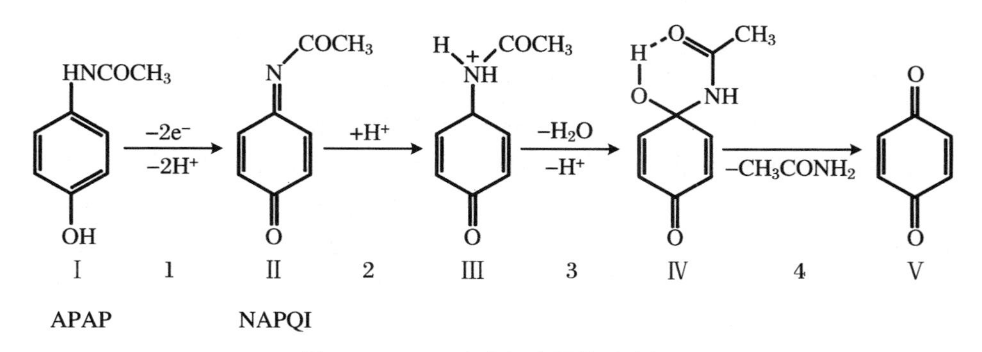

图3.3　APAP在电极表面的反应机理

上述机理可以通过改变溶液pH及扫描速度来证实。在pH＝6的缓冲溶液中，APAP在电极表面发生快速氧化反应，失去2个电子和2个质子，得到产物N-乙酰-对醌胺（NAPQI），如图3.3步骤1，由于反应过程中有H^+参与，因此APAP的氧化峰电位随溶液pH改变而发生变化。当溶液pH≥6时，NAPQI能以去质子化的形式稳定地存在于溶液中。因此在该pH范围内，APAP的循环伏安图中仅有一个氧化峰,没有还原峰。在一定条件下,该氧化峰电流与 APAP 的浓度呈线性相关,这是进行定量分析的依据。

在酸性条件下(如 pH=2.2),NAPQI 容易发生质子化反应生成产物Ⅲ,该产物稳定性不高,在快速扫描时可在循环伏安图中观察到一个还原峰。随后,产物Ⅲ会较快转化为产物Ⅳ,该产物在实验所采用的电位范围内不具有电活性。因此,仅在电位扫描速度较快时可以观察到产物Ⅲ的还原峰,若电位扫描速度较慢,则难以观察到产物Ⅲ的还原峰。在极酸条件下(如 1.8 mol/L 硫酸),产物Ⅳ可转化为苯醌(产物Ⅴ),所以在极高酸度条件下,除了产物Ⅲ的还原峰,还可以观察到苯醌的一个还原峰。

3.1.3 仪器与试剂

1. 仪器

CHI660E 电化学分析仪,三电极系统电解池,玻碳电极,铂电极,Ag-AgCl 参比电极。

2. 试剂

离子强度为 0.5,pH=2.2 和 pH=6.0 的 Mcllvaine 缓冲溶液,1.8 mol/L 硫酸溶液,0.03 mol/L 对乙酰氨基酚溶液,对乙酰氨基酚片,去离子水。

3.1.4 实验步骤

1. APAP 电氧化机理研究

(1) 配制 APAP 浓度为 3 mmol/L 的 pH=2.2、pH=6.0 的 Mcllvaine 缓冲溶液及 APAP 浓度为 3 mmol/L 的 1.8 mol/L 的硫酸溶液。

(2) 组装三电极体系,将电极引线接入电化学工作站。

(3) 在 3.0 mmol/L 的 pH=2.2、pH=6.0 的 Mcllvaine 缓冲溶液及 1.8 mol/L 的硫酸溶液中以 250 mV/s 的扫描速度进行扫描,记录循环伏安图。随后在 1.8 mol/L 硫酸的 3 mmol/L 的 APAP 溶液中以 250 mV/s,190 mV/s,140 mV/s,90 mV/s 和 40 mV/s 的扫描速度进行扫描,记录循环伏安图。观察在不同 pH 溶液中氧化峰和还原峰的情况。

2. 标准曲线法测定 APAP 浓度

(1) 配制浓度分别为 1 mmol/L,2 mmol/L,3 mmol/L,4 mmol/L 和 5 mmol/L 的对乙酰氨基酚标准溶液。

(2) 精密称取对乙酰氨基酚片粉 60 mg,置于 250 mL 烧瓶中,加入适量去离子水,加热煮沸至粉末溶解,稍冷后加入适量活性炭于烧瓶中,煮沸 5~10 min,趁热过滤,滤液冷却后有对乙酰氨基酚晶体析出,继续过滤,尽量除去母液,对晶体进行洗涤工作。随后取出晶体放在表面皿上晾干。将该晶体于60 ℃水浴40 min，放冷10 min，待溶解完全后转移至100 mL容量瓶中，加入去离子水定容，摇匀即得样品溶液。

(3) 采用循环伏安法测定不同浓度对乙酰氨基酚标准溶液，随后按照同样的操作步骤采用循环伏安法测定对乙酰氨基酚样品溶液，计算对乙酰氨基酚片中提取的对乙酰氨基酚含量。

3.1.5　数据记录与处理

1. 数据记录

(1) 扫描速度对氧化峰的影响(表3.1)。

表3.1

扫描速度 v(mV/s)	40	90	140	190	250
扫描速度平方根 $v^{1/2}$($(mV/s)^{1/2}$)					
氧化峰电流 I_p(μA)					

(2) pH对氧化峰的影响(表3.2)。

表3.2

pH	6.0	2.2	1.8 mol/L硫酸
氧化峰电位 E_p(mV)			

(3) APAP浓度和电流关系(表3.3)。

表3.3

APAP浓度 c(mmol/L)	1	2	3	4	5	样品
氧化峰电流 I_p(μA)						

2. 结果计算

(1) 以扫描速度平方根 $v^{1/2}$ 为横坐标，氧化峰电流 I_p 为纵坐标，绘制 $v^{1/2}$-I_p 关系曲线。

(2) 判断pH改变对 E_p 的影响。

(3) 根据循环伏安图,分别讨论 APAP 在不同 pH 溶液中的反应机理。

(4) 以溶液浓度 c 为横坐标,氧化峰电流 I_p 为横坐标,绘制 c-I_p 标准曲线,得到线性回归方程。

(5) 采用标准曲线法计算对乙酰氨基酚片中 APAP 的含量。

3.1.6 注意事项

(1) 每次做完实验均要仔细打磨电极,防止电极表面在实验过程中吸附其他物质,避免影响实验数据。

(2) 电极线与电极相连时要注意电极接口,避免相互接触发生短路。

(3) 溶液配制过程要操作规范,避免操作过程中的误差对实验结果产生影响,电极在使用前要彻底打磨干净。

3.1.7 思考题

(1) 如何通过实验结果验证图 3.3 步骤 1 中 APAP 的电化学氧化过程涉及两个电子和两个质子的反应?

(2) 实验步骤中改变扫描速度的目的是为了研究 APAP 电化学机理中的哪一步,理由是什么?

(3) 采用标准曲线法测定样品中 APAP 浓度时是否存在基体效应? 若存在,则采用什么定量方法更好?

(4) 伏安法测定试样中 APAP 时可能存在哪些干扰?

Chapter 3　Comprehensive Instrumental Analysis Experiments

3.1　Detection of Acetaminophen by Cyclic Voltammetry Method

3.1.1　Objectives

(1) Learn the principles of cyclic voltammetry.

(2) Master the process of electrochemical oxidation mechanism of acetaminophen by cyclic voltammetry.

(3) Learn to measure acetaminophen concentration in paracetamol by cyclic voltammetry.

3.1.2　Principles

Voltammetry is an electrochemical analysis method for qualitative or quantitative analysis based on the voltammetry characteristic curve. The variable potential applied to the electrodes is called the excitation signal. The four most common analytical methods in voltammetry include linear sweep voltammetry, differential pulse voltammetry, square wave voltammetry, and cyclic voltammetry. If the potential excitation signal is a triangular wave signal, the obtained voltammetry curve is the cyclic voltammetry curve. As shown in Figure 3.1, the scanning voltage is an isosceles triangle. In the process of voltage rise, if the compound is oxidized on the electrode surface, an oxidation reaction occurs. In the process of voltage reduction, if the oxidation product is reduced on the electrode surface, a reduction reaction occurs. Therefore, a triangular wave scan undergoes oxidation and reduction processes, so it is called cyclic voltammetry.

Redox peak potentials and their difference and redox peak currents are

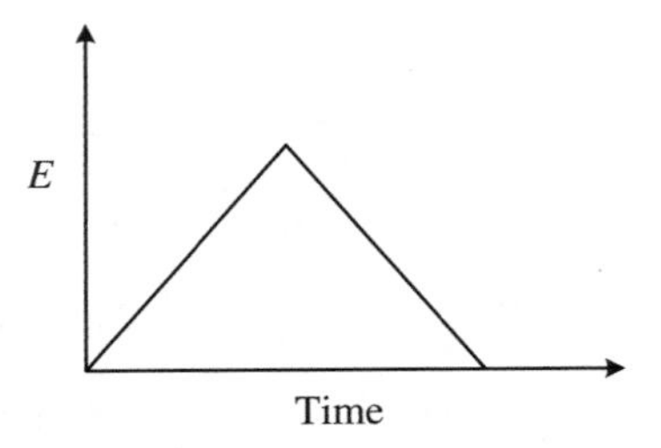

Figure 3.1 Time-voltage curve

important parameters in cyclic voltammetry. As shown in Figure 3.2, for a redox reaction, if the peak potential difference (ΔE_p) between the anodic peak potential (E_{ap}) and the cathodic peak potential (E_{cp}) is close to $0.059/n$ (V, 25 ℃), and the anodic peak current and cathodic peak current are approximately equal in value, the reaction is a reversible process. When the reaction is irreversible, the greater the difference in ΔE_p, the greater the degree of irreversibility.

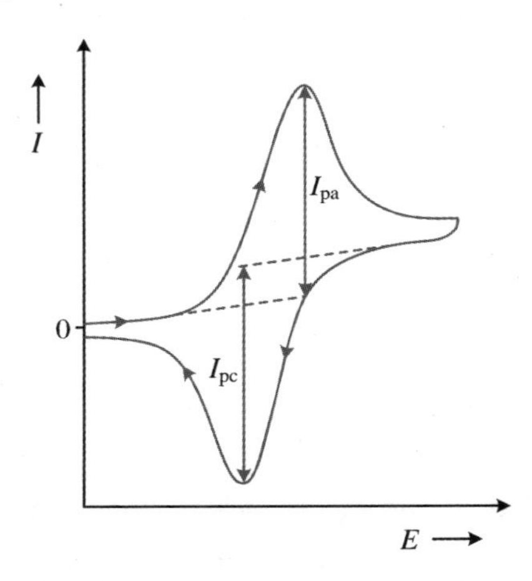

Figure 3.2 Current-voltage curve

Acetaminophen (APAP) is a nonsteroidal anti-inflammatory drug, which mainly acts as an antipyretic and analgesic by inhibiting the synthesis of prostaglandins. Cyclic voltammetry can be used to study the electron transfer process of electroactive substances on the electrode surface, which is an effective approach to study the reaction mechanism. Therefore, this experiment uses cyclic voltammetry to investigate the reaction mechanism of APAP on the electrode surface and the content of APAP in paracetamol. The reaction mechanism of APAP on the electrode surface can be expressed as Figure 3.3.

The above mechanism can be confirmed by changing the solution pH and

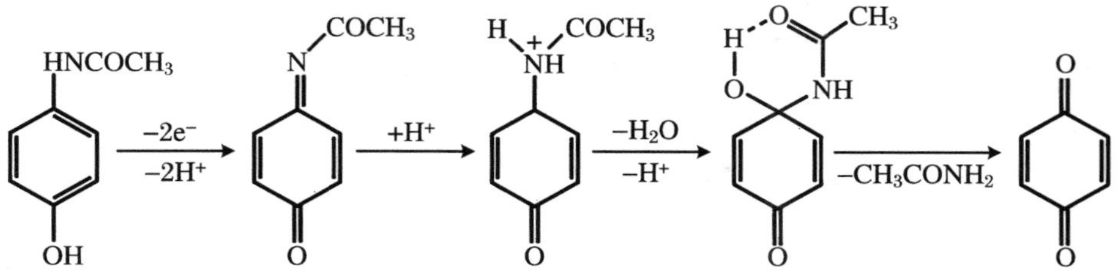

Figure 3.3　The reaction mechanism of APAP on the electrode surface

scanning rate. In the buffer solution (pH = 6.0), APAP undergoes a rapid oxidation reaction on the electrode surface, losing 2 electrons and 2 protons to obtain the product N-acetyl-p-quinoneamine (NAPQI), as shown in the first step in Figure 3.3. Due to the participation of H^+ in the reaction process, the oxidation peak potential of APAP changes with the pH of the solution. When the pH of the solution is greater than 6.0, NAPQI can stably exist in the solution in the deprotonated form. Therefore, in this pH range, there is only one oxidation peak and no reduction peak in the cyclic voltammogram of APAP. Under certain conditions, the oxidation peak current increases linearly with the concentration of APAP, which is the basis for quantitative analysis.

Under acidic conditions (eg. pH = 2.2), NAPQI is prone to protonation reaction to generate product Ⅲ. And a reduction peak can be observed in the cyclic voltammogram during fast scanning. Subsequently, product Ⅲ is rapidly converted to product Ⅳ, which is not electroactive in the potential range used in the experiments. Therefore, the reduction peak of product Ⅲ can be observed when the potential scanning rate is fast. However, when the potential scanning rate is slow, it is difficult to observe the reduction peak of product Ⅲ. Under extremely acidic conditions (such as 1.8 mol/L sulfuric acid), product Ⅳ can be converted into benzoquinone (product Ⅴ). Under extremely high acidity conditions, in addition to the reduction peak of product Ⅲ, a reduction of benzoquinone can also be observed.

3.1.3　Instruments and Reagents

1. Instruments

CHI660E electrochemical analyzer, three-electrode system electrolytic cell, glassy carbon electrode, platinum electrode, Ag-AgCl reference electrode.

2. Reagents

Mcllvaine buffer solution with ionic strength of 0.5, pH=2.2 and pH=6.0, 1.8 mol/L sulfuric acid solution, 0.03 mol/L acetaminophen solution, paracetamol tablets, deionized water.

3.1.4 Procedures

1. Electro-oxidation Mechanism of APAP

(1) Prepare the Mcllvaine buffer solution of pH=2.2 and pH=6.0 with APAP concentration of 3 mmol/L, and the sulfuric acid solution of 1.8 mol/L of APAP concentration of 3 mmol/L.

(2) Assemble the three-electrode system, and connect the electrode, leading to the electrochemical workstation.

(3) Record the cyclic voltammograms in 3.0 mmol/L Mcllvaine buffer (pH=2.2, pH=6.0) and acidic solution (1.8 mol/L sulfuric acid) at the scan rate of 250 mV/s. Subsequently, record the cyclic voltammograms by scanning at speed of 250 mV/s, 190 mV/s, 140 mV/s, 90 mV/s and 40 mV/s in sulfuric acid (1.8 mol/L) of APAP (3 mmol/L). Observe the oxidation and reduction peaks in different pH solutions.

2. Determination of APAP Concentration by Standard Curve Method

(1) Prepare acetaminophen standard solutions with concentrations of 1 mmol/L, 2 mmol/L, 3 mmol/L, 4 mmol/L and 5 mmol/L, respectively.

(2) Add 60 mg of paracetamol acetaminophen powder into a 250 mL flask with an appropriate amount of pure water. Then, heat and boil until the powder is dissolved. After cooling slightly, mix an appropriate amount of activated carbon to the flask, boil for 5~10 min, and filter while hot. After the filtrate has cooled, acetaminophen crystals can be precipitated. Then, the filtrate should continue to be filtered to remove the mother liquor. Subsequently, wash the crystals, remove and place on a watch glass to dry. The powder ought to be placed in a 60 ℃ water bath for 40 min, allowed to cool for 10 min, and then transfer to a 100 mL volumetric flask after complete dissolution. Add an appropriate amount of deionizer to the volume and shake to obtain the sample solution.

(3) Use cyclic voltammetry to determine the standard solutions of acetaminophen with different concentrations, and then extract the content of

acetaminophen from paracetamol determined by cyclic voltammetry according to the same operation process.

3.1.5 Data Recording and Processing

1. Data Recording

(1) The effect of scan rate on oxidation peak (Table 3.1).

Table 3.1

Scan rate v(mV/s)	40	90	140	190	250
Square root of scan speed $v^{1/2}$ ($(mV/s)^{1/2}$)					
Oxidation peak current I_p (μA)					

(2) The effect of pH on oxidation peak (Table 3.2).

Table 3.2

pH	6.0	2.2	1.8 mol/L sulfuric acid
Oxidation peak potential E_p (mV)			

(3) Relationship between APAP concentration and current (Table 3.3).

Table 3.3

APAP Concentration c (mmol/L)	1	2	3	4	5	Sample
Oxidation peak current I_p (μA)						

2. Result Calculation

(1) Taking the square root of the scanning speed $v^{1/2}$ as the abscissa and the oxidation peak current I_p as the ordinate, plot the curve of $v^{1/2}$-I_p.

(2) Evaluation of the effect of pH change on E_p.

(3) According to the cyclic voltammogram, discuss the reaction mechanism of

APAP in different pH solutions.

(4) Taking the solution concentration c as the abscissa and the oxidation peak current I_p as the abscissa, plot the curve of c-I_p, and obtain the linear regression equation.

(5) Calculate APAP content in paracetamol by standard curve method.

3.1.6 Cautions

(1) After each experiment, the electrode should be carefully polished to prevent the adsorption of other substances on the surface of the electrode during the experiment and avoid affecting the experimental data.

(2) When the electrode wire is connected to the electrode, pay the electrode interface should be restrained to contact with each other due to the short circuit.

(3) During the preparation of the experimental solution, the operation should be standardized to avoid errors in the operation process which can affect the experimental results. The electrodes should be thoroughly polished before use.

3.1.7 Questions

(1) How can you verify that the electrochemical oxidation process of APAP involves the reaction of two electrons and two protons in step 1 of Figure 3.3 through the experimental results?

(2) Which step in the electrochemical mechanism of APAP is the purpose of changing the scanning rate in the experimental steps to investigate, and why?

(3) Is there a matrix effect in the determination of APAP concentration in samples, standard curve method? If so, which quantitative method is better?

(4) What are the possible interferences in the determination of APAP in a sample by voltammetry?

3.2　双波长等吸收点法测定速洁舒洗剂中醋酸氯己定的含量

3.2.1　实验目的

（1）掌握双波长分光光度法的基本原理；掌握复方制剂不经分离直接测定各组分含量的方法。

（2）熟悉单波长型分光光度计进行双波长法测定的方法。

3.2.2　基本原理

1．波长的选择

在待测组分（Ⅰ）的最大吸收波长（测定波长，λ_1）处测定待测组分和干扰组分（Ⅱ）吸收度的总和；另选一适当波长（参比波长，λ_2）测定吸收度，并使干扰组分在测定波长和参比波长处的吸收度相等，即 $A_{\lambda_1}=A_{\lambda_2}$，而待测组分在这两个波长处吸收度的差值足够大（图3.4）。

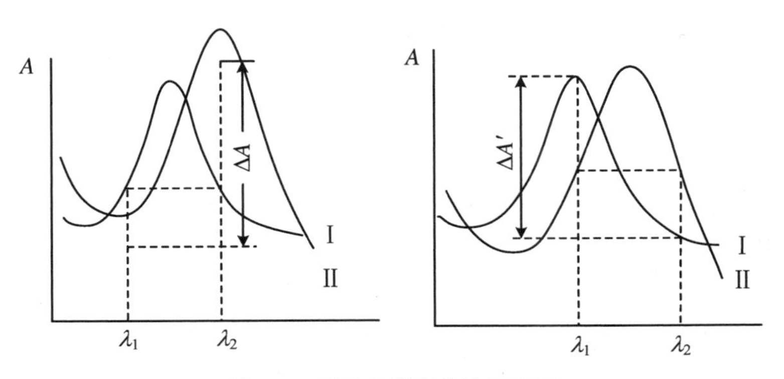

图3.4　双波长等吸收法示意图

2．定量依据

定量依据即样品在两个波长下吸收度的差值（ΔA）：吸收度差值（ΔA）仅与待测组分的浓度有关，而与干扰组分无关，干扰组分的干扰被消除。

3．基本条件

（1）干扰组分在两个波长处吸收度相等（$\Delta A=0$），以消除其干扰。

（2）待测组分在两个波长的 ΔA 足够大，以满足定量分析的要求。

（3）波长 λ_1、λ_2 应选在吸收曲线的平缓处，以减小测定误差。

4. 速洁舒洗液的紫外光谱

分别配制 10 mg·L^{-1}的替硝唑、醋酸氯己定溶液,以蒸馏水为空白进行紫外扫描,结果显示,醋酸氯己定在波长 253 nm 处有吸收峰,替硝唑在此有等吸收点,故可用双波长等吸收点法测定醋酸氯己定的含量(图 3.5)。

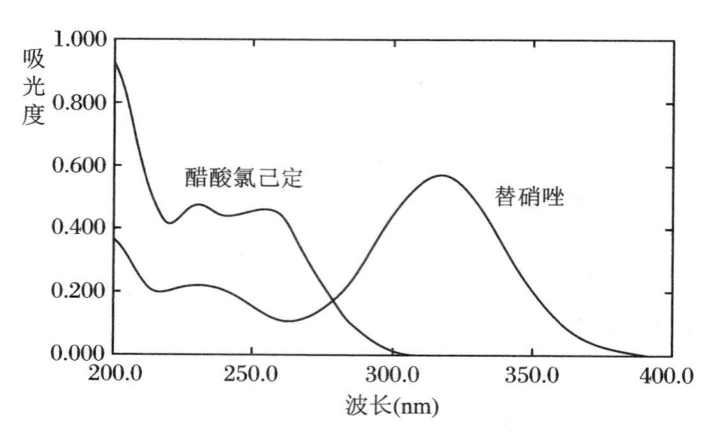

图 3.5 醋酸氯己定和替硝唑的吸收光谱

3.2.3 仪器及试剂

替硝唑、醋酸氯己定、紫外分光光度计。

3.2.4 实验步骤

速洁舒洗液的配制:称取替硝唑(TN)0.2 g,醋酸氯己定(CA)0.2 g,加蒸馏水至 1000 mL,混匀,即得。

1. 标准曲线的制作

取醋酸氯己定对照品 0.02 g,精密称定,溶于 100 mL 量瓶中并稀释至刻度,分别精密量取 2.0 mL、3.0 mL、3.5 mL、4.5 mL、5.0 mL 醋酸氯己定溶液于 50 mL 量瓶中并稀释至刻度,分别在波长 253 nm、273 nm 处测定吸收度,求 ΔA,并计算其回归方程。

2. 样品的测定

精密量取本品 2 mL 于 50mL 量瓶中,用蒸馏水稀释至刻度,按分光光度法于波长 253 nm,273 nm 处测吸光度。按回归方程计算含量。

3.2.5　数据记录与处理

根据所求得的回归方程计算醋酸氯己定的含量。

3.2.6　注意事项

当吸光度处在吸收曲线的陡然上升或下降的部位时，波长的微小变化可能对测定结果造成显著影响，故应注意测试条件的一致性。

3.2.7　思考题

双波长等吸收点法的基本条件及定量依据是什么？

3.2 Content Determination of Chlorhexidine Acetate in SuJieShu Detergent by Dual-wavelength Isoabsorption Point Method

3.2.1 Objectives

(1) Master the basic principle of double-wavelength spectrophotometry, and the method of directly measuring the content of each component without separation of compound preparations.

(2) Be familiar with a single-wavelength spectrophotometer for two-wavelength measurement.

3.2.2 Principles

1. Selection of Wavelength

The sum of the absorbance of the component to be measured and the interfering component (Ⅰ) must measured at the maximum absorption wavelength (measurement wavelength, λ_1) of the component to be measured (Ⅱ); another appropriate wavelength (reference wavelength, λ_2) should be selected to measure the absorbance and make the absorbance of the interfering component at the measurement wavelength and the reference wavelength equal, i.e., $A_{\lambda_1} = A_{\lambda_2}$ while the difference of the absorbance of the component to be measured at these two wavelengths is large enough (Figure 3.4).

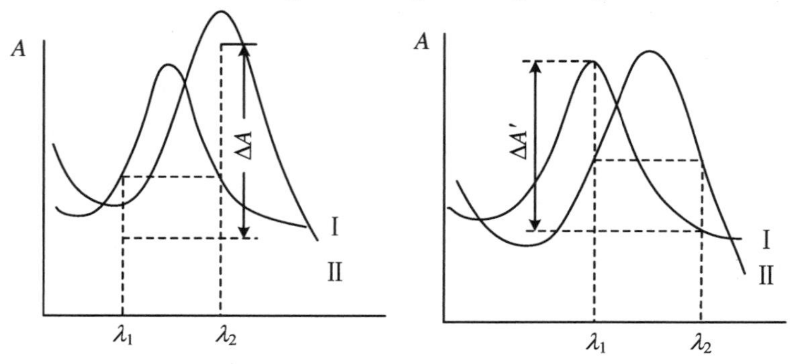

Figure 3.4 Dual-wavelength isoabsorption point method

2. Quantitative Basis

The quantitative basis is based on the difference in absorption of the two wavelengths (ΔA): the difference in absorption value (ΔA) is only related to the concentration of the component to be tested, but not to the interference component, which is eliminated.

3. Basic Conditions

(1) The interference component should have equal absorption at the two wavelengths ($\Delta A = 0$) to eliminate its interference.

(2) The ΔA of the component to be measured at the two wavelengths must be large enough to meet the requirements of quantitative analysis.

(3) The wavelengths λ_1 and λ_2 should be selected at the gentle level of the absorption curve to reduce the measurement error.

4. UV Spectra of the Detergent

Solution of tinidazole and chlorhexidine acetate must be prepared respectively, and use distilled water as blank for UV scanning. The results show that chlorhexidine acetate has absorption peak at 253 nm, and tinidazole has an equal absorption point here, so the content of chlorhexidine acetate could be determined by the dual wavelength equal absorption point method (Figure 3.5).

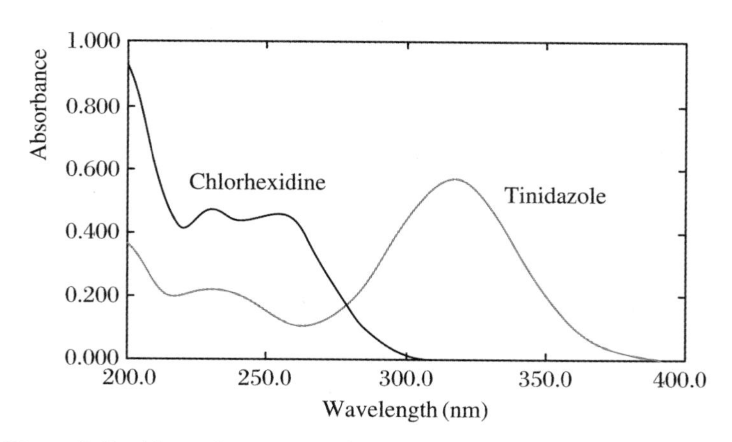

Figure 3.5　Absorption spectra of chlorhexidine acetate and tinidazole

3.2.3　Instruments and Reagents

Tinidazole, chlorhexidine acetate, and UV spectrophotometer.

3.2.4 Procedures

Preparationss of detergent: weigh 0.2 g tinidazole (TN), 0.2 g chlorhexidine acetate (CA), add distilled water to 1000 mL, mix them, and then it's prepared.

1. Production of the Standard Curve

Weigh 0.02 g of chlorhexidine acetate reference substance precisely, dissolve in 100 mL and dilute to the scale. Measure 2.0 mL, 3.0 mL, 3.5 mL, 4.5 mL, 5.0 mL precisely in 50 mL and dilute to the scale, measure the absorbance at 253 nm and 273 nm respectively, find ΔA, and calculate the regression equation.

2. Determination of Samples

Place 2 mL of the product in a 50 mL volumetric flask, dilute with distilled water to the scale and measure the absorbance at 253 nm and 273 nm according to the spectrophotometric method. Calculate the content according to the regression equation.

3.2.5 Data Recording and Processing

Calculate the content of chlorhexidine acetate according to the obtained regression equation.

3.2.6 Cautions

When the absorbance is in the steeply rising or falling part of the absorption curve, a small change in wavelength may have a significant effect on the measurement results, so attention should be paid to the consistency of the test conditions.

3.2.7 Questions

What are the basic conditions and quantitative basis of the dual wavelength equal absorption point method?

3.3　安钠咖注射液中苯甲酸钠和咖啡因的含量测定

3.3.1　实验目的

(1) 掌握和了解双波长分光光度法测定二元混合物的原理和方法。
(2) 掌握标准曲线的绘制和应用。
(3) 学习紫外可见分光度计的操作方法和结构。

3.3.2　基本原理

当光照射物质时,物质对光的吸收遵循朗伯-比尔定律,即当一定波长的入射光照射某物质的溶液时,溶液对光的吸光度 A 与物质的浓度以及溶液的厚度成正比关系:

$$A = Elc$$

式中 A 为吸光度,l 为液层厚度(单位:cm),c 为被测物质的浓度,E 为吸收系数。E 值随 c 所取单位不同而不同。如果浓度 c 以物质的量浓度(mol/L)表示,则 E 用 ε 表示,称为摩尔吸收系数。如果浓度 c 以质量百分浓度(g/100 mL)表示,则 E 用 $E_{1\mathrm{cm}}^{1\%}$ 表示,称为百分吸收系数。物质的摩尔吸光系数 ε 与波长和溶剂有关,在溶剂一定的情况下,只取决于波长 λ。

吸光度还具有加和性,即当溶液中存在两种或两种以上吸光物质时,只要共存物质彼此之间不发生相互作用,则测得的吸光度将是各物质吸光度的加和,便有以下关系:

$$A = A_1 + A_2 + \cdots + A_n$$

吸光度的加和性是分光光度法测定混合组分的定量依据。利用此性质可不经分离对混合物中的组分进行测定。

假设混合物中只含有 a 和 b 两种物质,a 物质、b 物质以及两者的混合物 c 的吸收光谱如图 3.6 所示。若 a 与 b 的吸收曲线重叠,则可采用双波长法中的等吸收点法进行测定。方法如下:首先在 a 物质的吸收曲线上选取点最大吸收波长 λ_1 对应的 O 点,再通过 O 点对 X 轴做垂直线,并与 b 物质的吸收曲线相交于 P 点。再从 P 点出发向 X 轴做水平线,并于 b 物质的吸收曲线相交于另一点 Q,Q 点对应的波长为 λ_2。假设在 1 cm 的吸收池中分别测定波长 λ_1 和 λ_2 处 a 和 b 物质的吸光度。由朗伯-比尔定律和吸光度的加和性得:

两物质在波长 λ_1 处的总吸光度

$$A_1 = A_{1(\mathrm{a})} + A_{1(\mathrm{b})} = E_{1(\mathrm{a})}c_{(\mathrm{a})} + A_{1(\mathrm{b})}$$

两物质在波长 λ_2 处的总吸光度

$$A_2 = A_{2(a)} + A_{2(b)} = E_{2(a)} c_{(a)} + A_{2(b)}$$

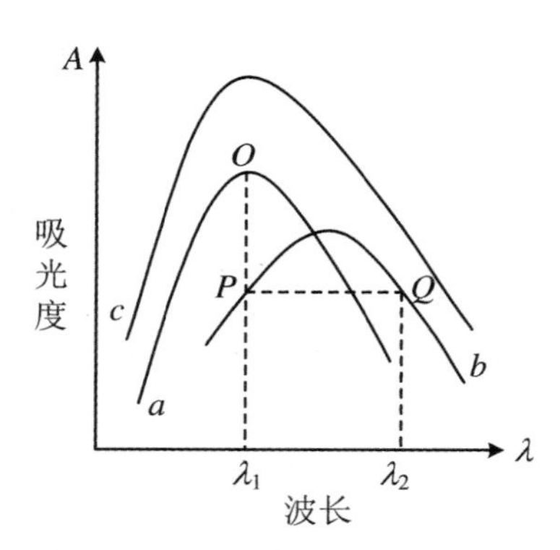

图 3.6 等吸收点法

由于 $A_{1(b)} = A_{2(b)}$,所以

$$\Delta A = A_2 - A_1 = E_{2(a)} c_a - E_{1(a)} c_a = (E_{2(a)} - E_{1(a)}) c_a$$

对于物质 a 来说,$E_{2(a)}$和 $E_{1(a)}$都是定值,因此 $\Delta A = kc_a$,且此 λ_2 与 λ_1 的吸光度之差只与物质 a 的浓度成正比,与 b 物质浓度无关,因此通过该方法就可测出物质 a 的浓度。同理,可选取物质 b 的最大吸收波长 λ_3 以及通过等吸收点方法找到的波长 λ_4 来测定物质 b 的浓度。

可用上述方法分别测定安钠咖注射液中的苯甲酸钠和咖啡因的含量。苯甲酸钠和咖啡因在 0.1 mol/L HCl 溶液分别在 230 nm 和 272 nm 处有最大吸收峰(图 3.7)。若测定苯甲酸钠含量,咖啡因会在 230 nm 和 258 nm 两处具有相等的吸光度。若测定咖啡因含量,苯甲酸钠会在 272 nm 和 254 nm 两处具有相等的吸光度。可选 230 nm 和 258 nm、272 nm 和 254 nm 四个波长作为测定波长分别测定苯甲酸钠和咖啡因的含量。应注意的是,在不同仪器上会出现不同的波长,应对波长组合进行校正。

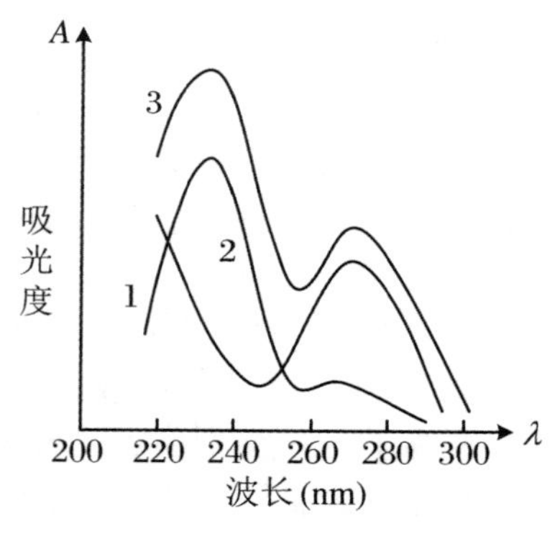

图 3.7 苯甲酸钠和咖啡因的吸收光谱

1. 咖啡因; 2. 苯甲酸钠; 3. 两者混合物

3.3.3　仪器与试剂

1. 仪器

752型紫外分光光度计、50 mL容量瓶、刻度吸管(5 mL)。

2. 试剂

咖啡因(对照品)、苯甲酸钠(对照品)、安钠咖注射液、0.1 mol/L HCl溶液。

3.3.4　实验步骤

1. 标准储备液的配制

为了配制0.1250 mg/mL的标准品溶液，需准确称取咖啡因和苯甲酸钠样品各0.03125 g，再分别用少量蒸馏水溶解并配成250 mL溶液。

2. 标准溶液的配制及吸收曲线的绘制

分别量取咖啡因和苯甲酸钠的标准储备液各3.00 mL，再分别加入50 mL的容量瓶中。再用0.1 mol/L的HCl溶液稀释至刻度，摇匀即得标准溶液。使用紫外可见分光光度计并在210～320 nm范围内自动扫描，得到两个物质的吸收曲线，再找出等吸收点。

3. 标准混合溶液的配制

配制一系列浓度的苯甲酸钠和咖啡因的标准混合溶液，分别量取苯甲酸钠和咖啡因的标准储备液1.00 mL、2.00 mL、3.00 mL、4.00 mL、5.00 mL至5只50 mL的容量瓶中，再使用0.1 mol/L的HCl溶液分别稀释至刻度，摇匀后即得浓度分别为2.5 μg/mL、5.0 μg/mL、7.5 μg/mL、10 μg/mL、12.5 μg/mL的咖啡因和苯甲酸钠的标准混合溶液。

4. 样品溶液的配制

在50 mL容量瓶中加入安钠咖注射液2.00 mL，并用蒸馏水稀释至刻度，并摇匀(标记为样品溶液Ⅰ)。再量取5.00 mL样品溶液Ⅰ于50 mL容量瓶中，并用蒸馏水稀释至刻度并摇匀(标记为样品溶液Ⅱ)。再量取5.00 mL样品溶液Ⅱ于50 mL容量瓶中，并用0.1 mol/L的HCl溶液稀释至刻度并摇匀，即得最终样品溶液。整个过程样品溶液被稀释了2500倍。

5. 测定

用紫外可见分光光度计分别在230 nm和258 nm、272 nm和254 nm处测定标准混合溶液的吸光度。再在上述四个波长下测定样品溶液的吸光度。

3.3.5　数据记录与处理

(1) 咖啡因和苯甲酸钠标准溶液在210～320 nm范围内的吸收曲线。

(2) 5个浓度的标准混合溶液以及样品溶液在230 nm和258 nm、272 nm和254 nm处的吸光度(表3.4)。

表3.4

c(μg/mL)	2.5	5.0	7.5	10.0	12.5	样品溶液
$A_{(230\ nm)}$						
$A_{(258\ nm)}$						
$A_{(254\ nm)}$						
$A_{(272\ nm)}$						
$\Delta A_{(苯甲酸钠)}$						
$\Delta A_{(咖啡因)}$						

(3) 绘制标准曲线。

① 苯甲酸钠的标准曲线:以c(μg/mL)为横坐标,$\Delta A_{(苯甲酸钠)}$为纵坐标。

② 咖啡因的标准曲线:以c(μg/mL)为横坐标,$\Delta A_{(咖啡因)}$为纵坐标。

(4) 计算安钠咖注射液中苯甲酸钠和咖啡因的含量。

3.3.6 注意事项

选取波长λ_1和λ_2时应注意两点:

(1) 尽量使物质a在λ_1或λ_2处有最大吸光度,这样可增大检测灵敏度。

(2) 物质b在波长λ_1和λ_2处必须吸光度相等,即有等吸收点。

3.3.7 思考题

(1) 如何根据吸收曲线选择等吸收点同时测定两种物质?

(2) 实验中的四个波长在吸收曲线中分别有什么意义?

3.3 Content Determination of Sodium Benzoate and Caffeine in Caffeine and Sodium Benzoate Injection

3.3.1 Objectives

(1) Understand and master the principles and methods of dual-wavelength spectrophotometry for the determination of binary mixtures.

(2) Master the drawing and application of standard curve.

(3) Learn the operation and structure of UV-Vis spectrophotometer.

3.3.2 Principles

When light irradiates a substance, the absorption of light by the substance follows the Lambert-Beer law. When a certain wavelength of incident light irradiates the solution of a substance, the absorbance (A) of the solution to light is proportional to the concentration of the substance and the thickness of the solution.

$$A = Elc$$

where A is the absorbance, l is the thickness of the liquid layer (unit: cm), c is the concentration of the tested substance, and E is the absorption coefficient. The value of E varies with the unit of c. If the concentration c is represented by the molar concentration of the substance (mol/L), E is represented by ε (the molar absorption coefficient). If the concentration c is expressed by mass percent concentration (g/100 mL), E is expressed by $E_{1\,\mathrm{cm}}^{1\%}$ (the percent absorption coefficient). The molar absorption coefficient of a substance is related to the wavelength and the solvent. For certain solvents, it only depends on the wavelength λ.

The absorbance is also additive. When there are two or more light-absorbing substances in the solution, the measured absorbance will be the sum of all absorbances of each substance when the coexisting substances do not interact with each other. The sum of absorbances obeys the following relation:

$$A = A_1 + A_2 + \cdots + A_n$$

The absorbance additivity is the quantitative basis for spectrophotometric determination of mixed components. This property allows the determination of components in a mixture without separation.

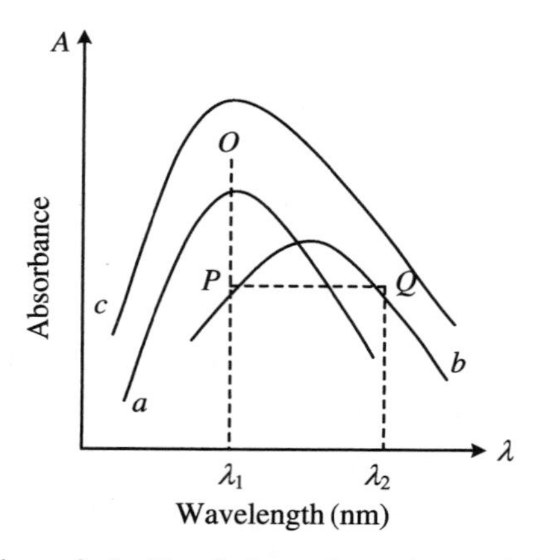

Figure 3.6 Equal absorption point method

Assuming that there are only two substances a and b in the mixture, the absorption spectra of substance a, substance b, and the mixture c of the two substances are shown in Figure 3. 6. If the absorption curves of a and b overlap, the isoabsorption point method in the dual-wavelength method can be used to measure the contents of a and b. Firstly, the O point is selected according to the maximum absorption wavelength λ_2 on the absorption curve of substance a. Then a vertical line to the X axis is drawn through the O point. The vertical line intersects the absorption curve of substance b at point P. Then, a horizontal line to the X axis is drawn from point P, and intersects the absorption curve of substance b at another point Q. The wavelength corresponding to point Q is λ_2. Assuming that the absorbance of substance a and b at wavelengths $\lambda_1$and λ_2 are measured in the 1 cm absorption cell, respectively. According to the Lambert-Beer law and the additivity of absorbance,

total absorbance of two substances at wavelength λ_1:

$$A_1 = A_{1(a)} + A_{1(b)} = E_{1(a)} c_{(a)} + A_{1(b)}$$

total absorbance of two substances at wavelength λ_2:

$$A_2 = A_{2(a)} + A_{2(b)} = E_{2(a)} c_{(a)} + A_{2(b)}$$

Because $A_{1(b)}$ is equal to $A_{2(b)}$,

$$\Delta A = A_2 - A_1 = E_{2(a)} c_a - E_{1(a)} c_a = (E_{2(a)} - E_{1(a)}) c_a$$

For substance a, $E_{2(a)}$ and $E_{1(a)}$ are both fixed values, so $\Delta A = kc_a$, and the difference between the absorbances of λ_2 and λ_1 is only proportional to the concentration of substance a and is independent of the concentration of substance b. Therefore, the concentration of substance a can be measured by this method. Similarly, the maximum absorption wavelength λ_3 of substance b and the wavelength λ_4 found by the isoabsorption point method can be selected to determine the concentration of substance b.

The above methods can be used to determine the content of caffeine and sodium benzoate in injection, respectively. Sodium benzoate and caffeine have maximal absorption at 230 nm and 272 nm in 0.1 mol/L HCl solution, respectively (Figure 3.7). When the concentration of sodium benzoate is measured, caffeine will have equal absorbance at both 230 nm and 258 nm. When the concentration of caffeine is measured, sodium benzoate will have equal absorbance at both 272 nm and 254 nm. Four wavelengths of 230 nm and 258 nm, 272 nm and 254 nm can be selected as the determination wavelengths to determine the content of sodium benzoate and caffeine respectively. It should be noted that different wavelengths will appear on different instruments, and the wavelength combination should be corrected.

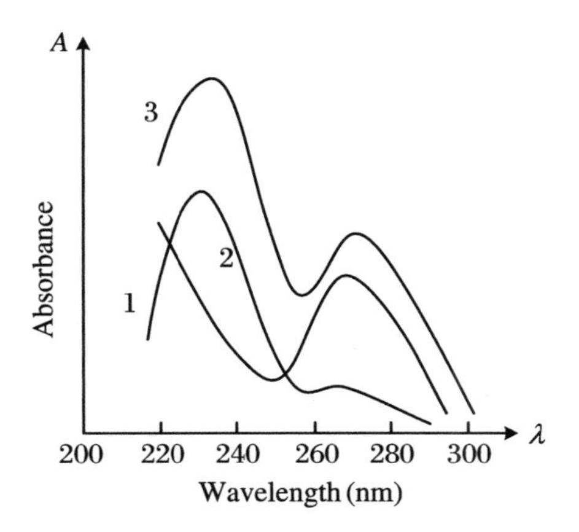

Figure 3.7　Absorption spectra of sodium benzoate and caffeine

1. Caffeine; 2. Sodium benzoate; 3. Mixture of caffeine and sodium benzoate

3.3.3　Instruments and Reagents

1. Instruments

752 UV spectrophotometer, volumetric flask (50 mL), graduated pipette (5 mL).

2. Reagents

Caffeine (reference substance), sodium benzoate (reference substance), ampulla injection, 0.1 mol/L HCl solution.

3.3.4 Procedures

1. Preparation of Standard Stock Solutions

To prepare a standard solution of 0.1250 mg/mL, 0.03125 g of caffeine and 0.03125 g of sodium benzoate are respectively dissolved in a small amount of distilled water to prepare the 250 mL solution.

2. Preparation of Standard Solution and Drawing of Absorption Curve

3.00 mL of the standard stock solutions of caffeine and sodium benzoate are respectively added to 50 mL volumetric flasks. The solutions should be diluted to the mark with 0.1 mol/L HCl solution and then shake to get the standard solution. The absorbance of the standard solutions is measured using UV-Vis spectrophotometer in the range of 210～320 nm. The isoabsorption points of caffeine and sodium benzoate are found from the absorption curves.

3. Preparation of Standard Mixed Solution

To prepare a series of standard mixed solutions of sodium benzoate and caffeine with different concentrations, 1.00 mL, 2.00 mL, 3.00 mL, 4.00 mL, 5.00 mL of standard stock solutions of sodium benzoate and caffeine are respectively transferred into five 50 mL volumetric flasks. These five solutions must be diluted to the mark using 0.1 mol/L HCl solution, resulting in the standard mixed solutions of caffeine and sodium benzoate with concentrations of 2.5 μg/mL, 5.0 μg/mL, 7.5 μg/mL, 10 μg/mL, and 12.5 μg/mL, respectively.

4. Preparation of the Sample Solution

2.00 mL of caffeine and sodium benzoate injection are added into a 50 mL volumetric flask, diluted to the mark with distilled water, and then shaken well. The resulting solution is marked as the sample solution Ⅰ. 5.00 mL of the sample solution Ⅰ is added into a 50 mL volumetric flask, diluted to the mark with distilled water, and is shaken well. The resulting solution is marked as the sample solution Ⅱ. Then 5.00 mL of the sample solution Ⅱ is added into a 50 mL volumetric flask, diluted to the mark with 0.1 mol/L HCl solution, and shaken well to obtain the final sample solution. During the whole process, the sample solution is diluted 2500 times in total.

5. Measurements

The absorbance values of the standard mixed solution and the sample solution are measured at 230 nm and 258 nm, 272 nm and 254 nm, respectively with a UV-visible spectrophotometer.

3.3.5 Data Recording and Processing

(1) Absorption curves of caffeine and sodium benzoate standard solutions in the range of 210～320 nm.

(2) The absorbance values of the standard mixed solution of five concentrations and the sample solution at 230 nm and 258 nm, 272 nm and 254 nm (Table 3.4).

Table 3.4

c(μg/mL)	2.5	5.0	7.5	10.0	12.5	Sample solution
$A_{(230\ nm)}$						
$A_{(258\ nm)}$						
$A_{(254\ nm)}$						
$A_{(272\ nm)}$						
$\Delta A_{(苯甲酸钠)}$						
$\Delta A_{(咖啡因)}$						

(3) Drawing of standard curve.

① Standard curve of sodium benzoate: take c(μg/mL) as the abscissa and ΔA (sodium benzoate) as the ordinate.

② The standard curve of caffeine: take c(μg/mL) as the abscissa and ΔA (caffeine) as the ordinate.

(4) Calculate the content of sodium benzoate and caffeine in the injection.

3.3.6 Cautions

Two points should be noted when selecting the wavelengths of λ_1 and λ_2:

(1) Try to make substance a have the maximum absorbance value at λ_1 or λ_2, which can increase detection sensitivity.

(2) Substance b must have the same absorbances at wavelengths of λ_1 and λ_2, that is, there are the isoabsorption points.

3.3.7 Questions

(1) How can we select the isoabsorption points according to the absorption curve to measure two substances at the same time?

(2) What is the significance of the four wavelengths in the experiment in the absorption curve?

3.4　气相色谱法测定祛伤消肿酊中樟脑、薄荷脑和龙脑的含量

3.4.1　实验目的

(1) 掌握内标法测定药物制剂中主成分含量的方法。
(2) 熟悉气相色谱仪的工作原理和操作方法。
(3) 了解气相色谱法在药物制剂含量测定中的应用。

3.4.2　基本原理

(1) 祛伤消肿酊是一种外用酊剂，收载于《中国药典》2020版(一部)，处方中含有多种挥发性成分：冰片、樟脑和薄荷脑。药典规定其每1 mL含樟脑($C_{10}H_{16}O$)应为24～36 mg；薄荷脑($C_{10}H_{20}O$)应为56～84 mg；含冰片以龙脑($C_{10}H_{18}O$)计，应不得少于26 mg。

(2)用内标法测定m克供试品中主成分含量时，精密称取m_i克对照品和m_s克内标物质，分别配成溶液，各精密量取适量，混合配成校正因子测定用的对照溶液。取一定量进样，记录色谱图。测量对照品和内标物质的峰面积或峰高，按下式计算相对校正因子：

$$\text{校正因子}(f') = \frac{m_i/A_i}{m_s/A_s}$$

式中，m_i为对照品物质的质量，A_i为对照品物质的峰面积或峰高，m_s为内标物质的质量，A_s为对内标物质的峰面积或峰高。

再取含有m_s克内标物质的供试品溶液，进样，记录色谱图，测量供试品中待测成分和内标物质的峰面积或峰高，按下式计算供试品中含有的待测成分含量：

$$\text{含量 } m_x(\%) = f' \times \frac{A_x}{m} \times 100\% = \frac{m_i/A_i}{m_s/A_s} \times \frac{A_x}{m} \times 100\%$$

式中，m_x表示供试品中待测组分的质量，A_x表示供试品中待测组分的峰面积或峰高。

3.4.3　仪器与试剂

1. 仪器

气相色谱仪(FID检测器)。

2. 试剂

无水乙醇(A. R.)。对照品：樟脑，龙脑，薄荷脑。内标物：萘(A. R.)。样品：祛伤消肿酊。

3.4.4 实验步骤

1. 色谱条件

色谱柱：以聚乙二醇 20000(PEG-20M)为固定相的毛细管柱(柱长为 30 m，内径为 0.32 mm，膜厚度为 0.25 μm)。理论板数按樟脑峰计算应不低于 5000。樟脑峰、薄荷脑峰、龙脑峰、内标物质峰彼此间的分离度应大于 1.5。柱温：130 ℃。进样口温度 220 ℃，检测器温度 250 ℃。载气(N_2)流速 3.0 mL/min，分流比 30∶1；尾吹气(N_2)流速 30 mL/min。

2. 溶液配制

(1) 对照品溶液的配制(测定校正因子用)。取萘适量，加无水乙醇溶解并稀释成每 1 mL 含 20 mg 的溶液，摇匀，作为内标溶液。另取樟脑对照品和龙脑对照品各约 10 mg，薄荷脑对照品约 20 mg，精密称定，同置 10 mL 量瓶中，精密加入内标溶液 1 mL，用无水乙醇稀释至刻度，摇匀。

(2) 供试品溶液的配制。精密量取待测样品 2 mL，置于 50 mL 量瓶中，精密加入内标溶液 5 mL，用无水乙醇稀释至刻度，摇匀，即得。

3. 样品测定

取对照品、供试品溶液各 1.0 μL，分别注入气相色谱仪，记录色谱图。

3.4.5 数据记录与处理

用内标校正因子法以色谱峰面积计算供试品中樟脑、薄荷脑、冰片(以龙脑计)的含量。

3.4.6 思考题

(1) 内标校正因子法进行含量测定适用于什么情况？

(2) 气相色谱法的内标如何选择？

3.4 Content Determination of Camphor, Menthol and Borneol in Qushang Xiaozhong Tincture by Gas Chromatography

3.4.1 Objectives

(1) Master internal standard method for the quantitation of the main components in pharmaceutical preparations.

(2) Be familiar with the working principle and operation of gas chromatograph.

(3) Understand the application of gas chromatography in the determination of the contents of pharmaceutical preparations.

3.4.2 Principles

(1) Qushang Xiaozhong Tincture (QXT) is a tincture for external use, which is listed in the 2020 edition of the *Chinese Pharmacopoeia* (Vol Ⅰ). The prescription of QXT contains several volatile components: borneol, camphor and menthol. The pharmacopoeia specified that each 1 mL of QXT should contain 24～36 mg of camphor ($C_{10}H_{16}O$), 56～84 mg of menthol ($C_{10}H_{18}O$) and not less than 26 mg (calculated using camphol ($C_{10}H_{18}O$)) of Borneolum Syntheticum.

(2) When the internal standard method is used for the quantitation of components in the test sample (m, g), the reference standard (m_i, g) and the internal standard (m_s, g) are accurately weighed (measured) and prepared into solutions, respectively. Appropriate amounts of each solution ought to be measured accurately and mixed to prepare a reference solution for the determination of the correction factor. A certain amount of the reference solution is injected in the chromatograph and the chromatogram is recorded. The peak area or peak height of the reference substance and the internal standard substance are measured. The relative correction factor is calculated as follows:

$$\text{Correction factor}(f') = \frac{m_i/A_i}{m_s/A_s}$$

Where m_i is the amount of the reference standard; A_i is the peak area or peak height of the reference standard; m_s is the amount of the internal standard; A_S is the peak area or peak height of the internal standard.

The test solutions containing the internal standard (m_s, g) are injected into the chromatograph. The peak area or peak height of each component is to be measured from chromatogram. The content is calculated according to the following formula:

$$\text{Content } m_x(\%) = f' \times \frac{A_x}{m} \times 100\% = \frac{m_i/A_i}{m_s/A_s} \times \frac{A_x}{m} \times 100\%$$

Where m_x is the amount of the analyte in the test solution; A_x is the peak area or peak height of the analyte in the test solution.

3.4.3 Instruments and Reagents

1. Instruments

Gas chromatograph (FID detector).

2. Reagents

Absolute ethanol (A. R.). Reference standard: camphor, borneol, and menthol. Internal standard: naphthalene (A. R.). Sample: Qushang Xiaozhong Tincture.

3.4.4 Procedures

1. Chromatographic Conditions

Chromatographic column: capillary column with polyethylene glycol 20000 (PEG-20M) as the stationary phase (30 m×0.32 mm, film thickness: 0.25 μm). The number of theoretical plates should not be lower than 5000 according to the camphor peak. The resolution between the peaks of camphor, menthol, borneol, and the internal standard should be greater than 1.5. Column temperature must be set at 130 ℃. The inlet temperature at 220 ℃, and the detector temperature at 250 ℃. The flow rates of the carrier gas (N_2) and the makeup gas (N_2) are 3.0 mL/min and 30 mL/min, respectively. The split ratio is 30 : 1.

2. Solution Preparation

(1) Preparation of the reference solution (for determination of correction

factor). Take an appropriate amount of naphthalene, add absolute ethanol to dissolve and dilute to a concentration of 20 mg/mL, shake well, and use it as the internal standard solution. In addition, transfer about 10 mg camphor and borneol reference standard respectively, weighed about 20 mg menthol reference standard accurately, into a 10 mL volumetric flask, add 1 mL of internal standard solution, dilute to volume with absolute ethanol, and shake well.

(2) Preparation of the test solution. Transfer 2.00 mL of the test sample, accurately measured, into a 50 mL volumetric flask, add precisely 5 mL of the internal standard solution, dilute to volume with absolute ethanol, and shake well.

3. Sample Determination

Inject 1.0 μL of the reference substance and the test solution respectively, into the gas chromatograph, and record the chromatograms.

3.4.5 Data Recording and Processing

The internal standard correction factor method is used to calculate the content of camphor, menthol and borneolum syntheticum (calculated as borneol) in the test sample based on the chromatographic peak area.

3.4.6 Questions

(1) What is the application of the internal standard correction factor method for content determination?

(2) How can we choose an internal standard for gas chromatography?

3.5 高效液相色谱法三种定量分析方法测定苯的含量

3.5.1 实验目的

(1) 掌握色谱法不同定量分析方法的特点。
(2) 熟悉不同定量分析方法对实验操作上要求的差异。
(3) 了解高效液相色谱法不同定量方法在应用上的选择。

3.5.2 基本原理

色谱法的定量依据是:在一定浓度范围内,色谱峰面积与组分的量成正比。利用反相色谱分离、紫外检测方法,在 254 nm 测定苯、联苯和菲各组分的峰面积,采用归一化法、外标法、内标法等常用定量分析方法对苯进行定量分析,并比较各定量分析方法的优缺点。

设样品质量为 m,待测组分的质量为 m_i,则其含量 P_i。

1. 归一化法

由

$$P_i = \frac{m_i}{m} \times 100$$

得

$$P_i = \frac{A_i f_i}{A_1 f_1 + A_2 f_2 + A_3 f_3 + \cdots + A_n f_n} \times 100$$

归一化法的特点:

(1) 无需准确控制进样量。
(2) 所有进样组分都出峰。

2. 内标法

在被分离物中加内标物,设其质量为 m_s,则有

$$\frac{m_i}{m_s} = \frac{A_i f_i}{A_s f_s}$$

$$P_i(\%) = \frac{m_i}{m} \times 100\% = \frac{A_i f_i}{A_s f_s} \times \frac{m_s}{m} \times 100\%$$

内标法的特点:

(1) 无需全部组分出峰,适合于常量成分中微量组分的定量。

(2) 内标物为样品中不含有的成分，保留时间与待测组分相近，$R>1.5$。

(3) 要准确称量内标物和样品的质量。

(4) 无需准确控制进样量。

3. 内标对比法

当内标法中校正因子未知时，在标准品 m_0 和样品 m_x 中分别加入内标物 $m_{s(0)}$ 和 $m_{s(i)}$，则

$$\frac{m_i}{m_{s(i)}} = \frac{A_i f_i}{A_{s(i)} f_s} \quad ①$$

$$\frac{m_0}{m_{s(0)}} = \frac{A_0 f_i}{A_{s(0)} f_s} \quad ②$$

①÷②得：

$$\frac{m_i/m_{s(i)}}{m_0/m_{s(0)}} = \frac{A_i f_i/A_{s(i)} f_s}{A_0 f_i/A_{s(0)} f_s} = \frac{A_i/A_{s(i)}}{A_0/A_{s(0)}}$$

$$P_i = \frac{m_i}{m_x} \times 100 = \frac{A_i/A_{s(i)}}{A_0/A_{s(0)}} \times \frac{m_{s(i)}}{m_{s(0)}} \times \frac{m_0}{m_x} \times 100$$

在标准品和样品中加入的内标物等量，则有

$$P_i = \frac{m_i}{m_x} \times 100 = \frac{A_i/A_{s(i)}}{A_0/A_{s(0)}} \times \frac{m_0}{m_x} \times 100$$

4. 外标法

又称标准曲线法，即以 A 为纵坐标，m 为横坐标，绘制标准曲线或回归方程，通过作图或计算得到样品中待测组分的含量。若标准曲线通过原点，可用外标点法；若标准曲线不通过原点，则常用外标两点法。

外标法的特点：

(1) 简单，不需要知道校正因子。

(2) 对仪器条件稳定性和操作要求高。

3.5.3 仪器与试剂

1. 仪器

高效液相色谱仪(色谱柱：C_{18}反相键合色谱柱，检测器：UV)，超声波清洗器，溶剂过滤器，石英亚沸高纯水蒸馏水器，微量注射器。

2. 试剂

甲醇，苯，联苯，菲(均为色谱纯)。

3.5.4 实验步骤

(1) 仪器操作。

色谱条件：流动相为甲醇：水(80：20)；流速为 1 mL/min；固定相为 C_{18}反相

键合色谱柱;检测波长为 254 nm;进样量为 20 μL (定量环)。

(2) 流动相处理:配制甲醇:水(80:20)并过滤、超声脱气。

(3) 苯标准储备液的制备:精确称取苯试剂 100 mg,置于 100 mL 容量瓶中,用甲醇稀释,定容至刻度,摇匀。

(4) 苯标准溶液的制备(200 μg/mL):精确吸取 20 mL 苯标准储备液,置于 100 mL 容量瓶中,用甲醇稀释,定容至刻度,摇匀。

(5) 联苯标准储备液的制备:精确称取联苯试剂 1 mg,置于 100 mL 容量瓶中,用甲醇稀释,定容至刻度,摇匀。

(6) 联苯标准溶液的制备(2 μg/mL):精确吸取 20 mL 联苯标准储备液,置于 100 mL 容量瓶中,用甲醇稀释,定容至刻度,摇匀。

(7) 菲标准储备液的制备:精确称取菲试剂 1 mg,置于 100 mL 容量瓶中,用甲醇稀释,定容至刻度,摇匀。

(8) 菲标准溶液的制备(1 μg/mL):精确吸取 10 mL 菲标准储备液,置于 100 mL 容量瓶中,用甲醇稀释,定容至刻度,摇匀。

(9) 萘标准储备液的制备:精确称取萘试剂 15 mg,置于 100 mL 容量瓶中,用甲醇稀释,定容至刻度,摇匀。

(10) 萘标准溶液的制备(15 μg/mL):精确吸取 10 mL 萘标准储备液,置于 100 mL 容量瓶中,用甲醇稀释,定容至刻度,摇匀。

(11) 苯和萘的混合溶液的制备(萘为内标):精确吸取 20 mL 苯标准储备液和 10 mL 萘标准储备液,置于 100 mL 容量瓶中,用甲醇稀释,定容至刻度,摇匀。

(12) 苯、联苯和菲的混合溶液的制备:精确吸取 20 mL 苯标准储备液、20 mL 联苯标准储备液以及 10 mL 菲标准储备液,置于 100 mL 容量瓶中,用甲醇稀释,定容至刻度,摇匀。

(13) 苯、联苯、菲和萘的混合溶液的制备(萘为内标):精确吸取 20 mL 苯标准储备液、20 mL 联苯标准储备液、10 mL 菲标准储备液以及 10 mL 萘标准储备液,置于 100 mL 容量瓶中,用甲醇稀释,定容至刻度,摇匀。

(14) 测定:吸取样品溶液 20 μL,注入色谱仪,记录色谱图,按归一化法、外标法、内标法计算苯的百分含量。

3.5.5 数据记录与处理

(1) 将实验数据计入表 3.5。

表 3.5

<table>
<tr><th colspan="3"></th><th>保留时间 t_R(min)</th><th>峰面积</th></tr>
<tr><td rowspan="6">对照品</td><td colspan="2">苯</td><td></td><td></td></tr>
<tr><td colspan="2">联苯</td><td></td><td></td></tr>
<tr><td colspan="2">菲</td><td></td><td></td></tr>
<tr><td colspan="2">萘</td><td></td><td></td></tr>
<tr><td rowspan="2">苯+萘
(二混)</td><td>苯</td><td></td><td></td></tr>
<tr><td>萘</td><td></td><td></td></tr>
<tr><td rowspan="7">供试品</td><td rowspan="3">苯+联苯+菲
(三混)</td><td>苯</td><td></td><td></td></tr>
<tr><td>联苯</td><td></td><td></td></tr>
<tr><td>菲</td><td></td><td></td></tr>
<tr><td rowspan="4">苯+联苯+菲+萘
(四混)</td><td>苯</td><td></td><td></td></tr>
<tr><td>联苯</td><td></td><td></td></tr>
<tr><td>菲</td><td></td><td></td></tr>
<tr><td>萘</td><td></td><td></td></tr>
</table>

(2) 数据处理:

苯、联苯和菲的混合物用于归一化法和外标法,苯和萘的混合物以及苯、联苯、菲、萘的混合物用于内标法。

3.5.6 注意事项

(1) 流动相应使用色谱纯试剂,水应为纯水及重蒸馏水,使用前应用 0.45 μm 的滤膜过滤和脱气(可在超声波清洗器上超声处理)。

(2) 进样前要排除微量注射器中的气泡。

(3) 柱塞清洗液为含 20%甲醇的水溶液,配制后抽滤,可循环使用。一般在夏季每半个月换一次,冬季可每月换一次。注意应保持管路及液体的清洁。

3.5.7 思考题

(1) HPLC 的三种定量分析方法各有什么特点?

(2) 试比较三种定量分析方法所得结果的差异,并分析原因。

3.5 Content Determination of Benzene by Three Quantitative Analysis Methods in High Performance Liquid Chromatography

3.5.1 Objectives

(1) Master the characteristics of different quantitative analysis methods of chromatography.

(2) Be familiar with the differences in experimental operation requirements of different quantitative analysis methods.

(3) Understand the application selection of different quantitative methods of high-performance liquid chromatography.

3.5.2 Principles

The quantitative basis for chromatography is that the peak area of the chromatography is proportional to the amount of components under a certain conditions. The peak area of benzene, biphenyl and phenanthrene components is determined at 254 nm by reversed-phase chromatography separation and ultraviolet detection method. Further, benzene is quantitatively analyzed by common quantitative analysis methods such as normalization method, external standard method, and internal standard method. The advantages and disadvantages of each quantitative analysis method are compared.

Assuming that the sample mass is m, and the mass of the component to be measured is m_i, its content P_i is:

1. Normalization Method

According to the formula $P_i = m_i/m \times 100$, get the following formula:

$$P_i = \frac{A_i f_i}{A_1 f_1 + A_2 f_2 + A_3 f_3 + \cdots + A_n f_n} \times 100$$

Features:

(1) There is no need to accurately control the injection volume.

(2) All incoming components are peaked.

2. Internal Standard Method

Add the internal standard to the separated substance, and set its mass as m_s, and obtain the following formula relationship:

$$\frac{m_i}{m_s}=\frac{A_i f_i}{A_s f_s}$$

$$P_i(\%)=\frac{m_i}{m}\times 100\%=\frac{A_i f_i}{A_s f_s}\times\frac{m_s}{m}\times 100$$

Features:

(1) It is not necessary to peak all components, and it is suitable for the quantification of trace components in constant components.

(2) The internal standard substance is not required to be contained in the sample, and the retention time is similar to that of the component being measured, $R>1.5$。

(3) The mass of the internal subject and sample should be accurately weighed.

(4) There is no need to accurately control the injection volume.

3. Internal Standard Comparison Method

When the correction factor is unknown in the internal standard method, add the internal standard $m_{s(0)}$ and $m_{s(i)}$ to the standard m_0 and sample m_x respectively, obtaining the following formula:

$$\frac{m_i}{m_{s(i)}}=\frac{A_i f_i}{A_{s(i)} f_s} \quad ①$$

$$\frac{m_0}{m_{s(0)}}=\frac{A_0 f_i}{A_{s(0)} f_s} \quad ②$$

①÷②:

$$\frac{m_i/m_{s(i)}}{m_0/m_{s(0)}}=\frac{A_i f_i/A_{s(i)} f_s}{A_0 f_i/A_{s(0)} f_s}=\frac{A_i/A_{s(i)}}{A_0/A_{s(0)}}$$

$$P_i=\frac{m_i}{m_x}\times 100=\frac{A_i/A_{s(i)}}{A_0/A_{s(0)}}\times\frac{m_{s(i)}}{m_{s(0)}}\times\frac{m_0}{m_x}\times 100$$

When the same amount of internal standard is added to the standard and the sample, the following formula is obtained:

$$P_i=\frac{m_i}{m_x}\times 100=\frac{A_i/A_{s(i)}}{A_0/A_{s(0)}}\times\frac{m_0}{m_x}\times 100$$

4. External Standard Method

Also known as standard curve method, the standard curve or regression equation is obtained with A as the vertical coordinate and m as the abscissa. Then the content of the components to be measured in the sample by drawing or calculating. If the standard curve passes through the origin, the external punctuation method can be used. If the standard curve does not pass the origin, the two-point method of external labeling is often used.

Features:

(1) This method is simple, the correction factor is not needed.

(2) It has high requirements for instrument condition stability and operation.

3.5.3 Instruments and Reagents

1. Instruments

High performance liquid chromatograph, (column is C_{18} reversed-phase bonded column, with UV detector), ultrasonic cleaner, solvent filter, quartz second boiling high pure water distiller, microsyringe.

2. Reagents

Methanol, benzene, biphenyl, phenanthrene (all chromatographically pure).

3.5.4 Procedures

(1) Instrument operation.

Chromatographic condition: mobile phase is methanol : water (80 : 20). flow rate is 1 mL/min; fixed phase is C_{18} reversed-phase bonded column; detection wavelength is 254 nm; injection volume is 20 μL (quantitative ring).

(2) Mobile phase treatment: prepare methanol : water (80 : 20) and filter, then ultrasonic degas.

(3) Preparation of benzene standard stock solution: weigh 100 mg benzene precisely and transfer into a 100 mL volumetric flask, then are diluted with methanol to the scale mark and shake well.

(4) Preparation of benzene standard solution (200 μg/mL): Place 20 mL benzene standard stock solution into a 100 mL volumetric flask, then dilute with methanol to the scale mark and shake well.

(5) Preparation of biphenyl standard stock solution: weigh 1 mg biphenyl precisely and transfer into a 100 mL volumetric flask, then are diluted with methanol to the scale mark and shake well.

(6) Preparation of biphenyl standard solution (2 μg/mL): Place 20 mL biphenyl standard stock solution into a 100 mL volumetric flask, then dilute with methanol to the scale mark and shake well.

(7) Preparation of phenanthrene standard stock solution: weigh 1 mg phenanthrene precisely and transfer into a 100 mL volumetric flask, then are diluted with methanol to the scale mark and shake well.

(8) Preparation of phenanthrene standard solution (1 μg/mL): Place 10 mL phenanthrene standard stock solution into a 100 mL volumetric flask, then dilute with methanol to the scale mark and shake well.

(9) Preparation of naphthalene standard stock solution: weigh 15 mg naphthalene precisely and transfer into a 100 mL volumetric flask, then are diluted with methanol to the scale mark and shake well.

(10) Preparation of naphthalene standard solution (15 μg/mL): Place 10 mL naphthalene standard stock solution into a 100 mL volumetric flask, then dilute with methanol to the scale mark and shake well.

(11) Preparation of mixed standard solution of benzene and naphthalene: place 20 mL benzene and 10 mL naphthalene standard stock solution into a 100 mL volumetric flask, then dilute with methanol to the scale mark and shake well.

(12) Preparation of mixed standard solution of benzene, biphenyl, and phenanthrene: place 20 mL benzene, 20 mL biphenyl, and 10 mL phenanthrene standard stock solution into a 100 mL volumetric flask, then dilute with methanol to the scale mark and shake well.

(13) Preparation of mixed standard solution of benzene, biphenyl, phenanthrene, and naphthalene: place 20 mL benzene, 20 mL biphenyl, 10 mL phenanthrene, and 10 mL naphthalene standard stock solution into a 100 mL volumetric flask, then dilute with methanol to the scale mark and shake well.

(14) Determination: 20 pL of sample solution is aspirated and injected into the chromatograph. The percentage content of benzene is calculated according to the normalization method, external standard method and internal standard method.

3.5.5 Data Recording and Processing

(1) Record the experimental data in Table 3.5.

表 3.5

			Retention time t_R(min)	Peak area
Reference substance	Benzene			
	Biphenyl			
	Phenanthrene			
	Naphthalene			
	Benzene + Naphthalene (two mixtures)	Benzene		
		Naphthalene		
Sample for test	Benzene + Biphenyl + Phenanthrene (three mixtures)	Benzene		
		Biphenyl		
		Phenanthrene		
	Benzene + Biphenyl + Phenanthrene + Naphthalene (Four mixtures)	Benzene		
		Biphenyl		
		Phenanthrene		
		Naphthalene		

(2) Data Processing:

Mixtures of benzene, biphenyls and phenanthrene are used for normalization and external standard methods. Mixtures of benzene and naphthalene and mixtures of benzene, biphenyls, phenanthrene, and naphthalene are used in the internal standard method.

3.5.6 Cautions

(1) Flow phase should be chromatographical pure reagents. Water should be pure water and distilled water, which should be filtered and degassed with a 0.45 μm filter membrane before use (can be sonicated on an ultrasonic cleaner).

(2) The bubbles in microsyringe should be eliminated before injection.

(3) The plunger cleaning liquid is an aqueous solution containing 20% methanol, which should also be filtered after preparation and can be recycled. Generally,

it should be changed every two weeks in summer and once a month in winter. The pipeline and liquid should keep clean.

3.5.7 Questions

(1) What are the characteristics of the three quantitative analysis methods of HPLC?

(2) Try to compare the differences in the results obtained by the three quantitative analysis methods and analyze the reasons.

3.6 高效液相色谱-质谱联用法鉴定双黄连口服液中的有效成分

3.6.1 实验目的

(1) 了解高效液相色谱-质谱联用仪的基本工作原理。
(2) 学习高效液相色谱-质谱联用仪的主要构造和基本操作方法。
(3) 掌握高效液相色谱-质谱联用的选择离子监测分析方法。

3.6.2 基本原理

高效液相色谱-质谱法(high performance liquid chromatography-mass spectrometry, HPLC-MS)是以液相色谱作为分离系统,质谱为检测系统的一个综合性分析技术。HPLC-MS体现了色谱和质谱优势的互补,将色谱对复杂样品的高分离能力,与MS具有高选择性、高灵敏度及能够提供相对分子质量与结构信息的优点结合起来,在药物分析、食品分析和环境分析等许多领域得到了广泛的应用。

HPLC-MS主要由液相色谱系统、连接接口、质量分析器和计算机数据处理系统组成。其主要过程为试样通过液相色谱系统进样,由色谱柱分离。从色谱仪流出的被分离组分依次通过接口进入质谱仪的离子源处并被离子化,然后离子被聚焦于质量分析器中,根据质荷比而分离,分离后的离子信号被转变为电信号,由电子倍增器进行检测,其检测信号被放大并传输到数据处理系统。

双黄连口服液由金银花、黄芩和连翘三味中药组成,具有良好的疏风解表、清热解毒的功效。绿原酸、咖啡酸、黄芩苷和木犀草素是双黄连口服液的主要活性成分,其结构如图3.8所示。本实验采用HPLC-MS鉴定中药双黄连口服液中绿原酸、咖啡酸、黄芩苷和木犀草素这四种有效成分。

3.6.3 仪器与试剂

1. 仪器

高效液相色谱-质谱联用仪(1100LC/MS Trap SL型,Agilent公司)。

2. 试剂

甲醇(色谱纯),甲酸(分析纯),超纯水。样品:双黄连口服液(市售)。对照品:绿原酸、咖啡酸、黄芩苷和木樨草素。

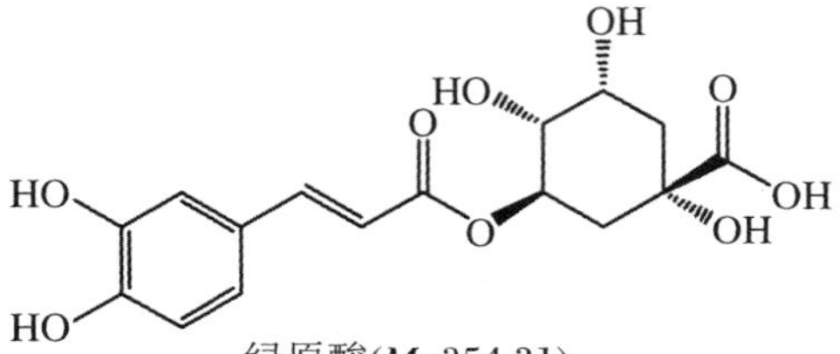

绿原酸(M=354.31) 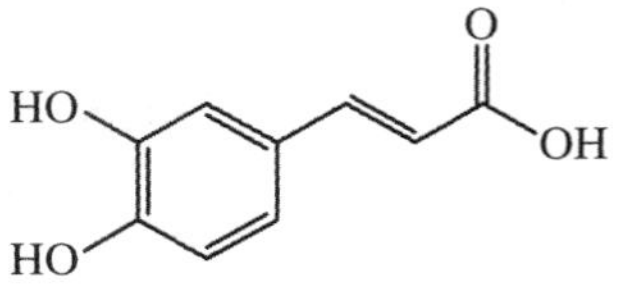　　咖啡酸(M=180.15)

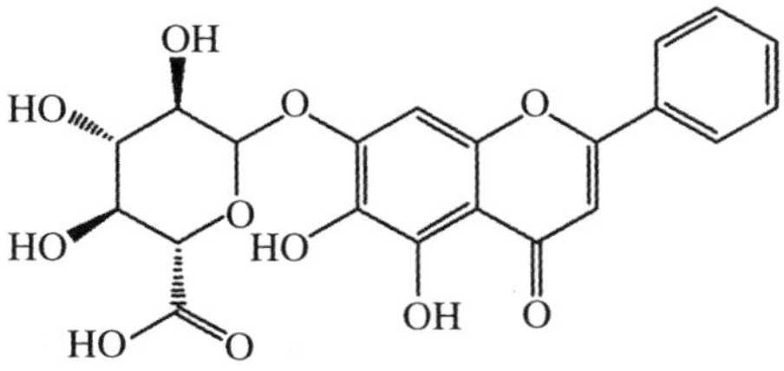

黄芩苷(M=446.37) 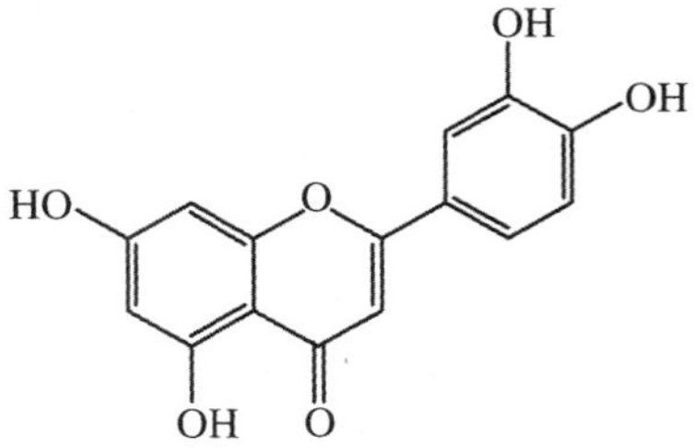　　本犀草素(M=286.24)

图3.8　绿原酸、咖啡酸、黄芩苷和木犀草素的结构示意图

3.6.4　实验步骤

1. 溶液的制备

(1) 对照品溶液的制备。分别精密称取适量绿原酸、咖啡酸、黄芩苷和木犀草素，用甲醇溶解定容，均配制成浓度10.0 μg/mL的溶液。分别量取4种对照品溶液各1 mL，混合制备成对照品的混合溶液。

(2) 样品溶液的制备。取双黄连口服液100.0 μL，用甲醇稀释并定容至10 mL，并使用0. 2 μm的微孔滤膜过滤，滤液供HPLC-MS分析。

2. 仪器操作条件(参考值)

(1) 色谱条件。

色谱柱：Johnson Spherigel C_{18}(250 mm×4.6 mm i.d.，粒径5 μm)；流动相：含0.3%甲酸的乙酸铵溶液(0.4 mmol /L)(A)-甲醇(B)；梯度洗脱：0～10 min(35% B)，11～25 min(65% B)，26～30 min(35% B)；流速：0.80 mL/min；柱温：25 ℃；进样量：20 μL。

(2) 质谱条件。

分流比3∶1，仅约0.2 mL/min进入质谱；ESI离子源，温度110 ℃；毛细管电压：4.0 kV，锥孔电压：25 kV；雾化气(N_2)和脱溶剂气(鞘气，N_2)流速分别为50 L/h和300 L/h，鞘气温度：300 ℃。ESI正离子检测模式，分时段选择离子模式(SIM)：0～7 min，m/z 377.4；7.0～12 min，m/z 181.0；12～18 min，m/z 447.1；18～25 min，m/z 287.1。

3. 样品测定

(1) 分别测定绿原酸、咖啡酸、黄芩苷和木犀草素的对照品溶液,得到色谱质谱图,其中分别选择 m/z 为 377.4(绿原酸的[M+Na]$^+$离子峰)、m/z 为 181.0(咖啡酸的[M+H]$^+$离子峰)、m/z 为 447.1(黄芩苷的[M+H]$^+$离子峰)和 m/z 为 287.1(木樨草素的[M+H]$^+$离子峰)的离子进行分段监测。

(2) 测定绿原酸、咖啡酸、黄芩苷和木犀草素对照品混合液。

(3) 测定样品,根据样品与对照品的峰面积比,采用外标对比法进行定量分析。

3.6.5 数据记录与处理

(1) 对样品中绿原酸、咖啡酸、黄芩苷和木犀草素色谱和质谱峰进行归属,并用分离度判断分离效果。

(2) 用外标对比法计算绿原酸、咖啡酸、黄芩苷和木犀草素的含量。

3.6.6 注意事项

(1) 流动相中含有的挥发性物质有利于液相组分在质谱的离子源中转化为气相离子。而非挥发性盐类(如磷酸盐缓冲液或离子对试剂)则不利于液相组分在质谱的离子源中转化为气相离子。因此 HPLC-MS 的流动相不能含有非挥发性盐。

(2) 使用 HPLC-MS 的正离子检测模式时除了出现组分的[M+H]$^+$离子峰,还会经常出现[M+Na]$^+$、[M+K]$^+$等形式的离子峰。各种形式的离子峰的质谱信号强度受实验条件的影响较大。

3.6.7 思考题

(1) HPLC-MS 与 MS 相比,在药物分析应用中的优越性主要体现在哪几个方面?

(2) 影响 HPLC-MS 质谱信号强度的主要因素有哪些?

(3) 是否可以用质谱的负离子模式检测本实验中的分析对象?如可以,是以何种离子峰出现?

3.6 Identification of Active Components in Shuanghuanglian Oral Liquid by High Performance Liquid Chromatography-Mass Spectrometry

3.6.1 Objectives

(1) Understand the basic working principle of high-performance liquid chromatography-mass spectrometry (HPLC-MS).

(2) Learn the main structure and basic operation methods of HPLC-MS.

(3) Master the selected ion monitoring and analysis method of HPLC-MS.

3.6.2 Principles

HPLC-MS is a comprehensive analysis technique with liquid chromatography as the separation system and mass spectrometry as the detection system. HPLC-MS reflects the complementarity of the advantages of chromatography and mass spectrometry. It combines the high separation ability of chromatography for complex samples with the advantages of high selectivity, high sensitivity, and the ability to provide relative molecular mass and structural information of MS. It is widely used in many fields such as pharmaceutical analysis, food analysis and environmental analysis.

HPLC-MS is mainly composed of liquid chromatography system, connection interface, mass analyzer and computer data processing system. The main process is that the sample would be injected through the liquid chromatography system and separated in the chromatographic column. The separated components flowing out of the chromatograph sequentially enter the ion source of the mass spectrometer through the interface and are ionized, and then the ions are focused in the mass analyzer, separated according to the mass-to-charge ratio, and the separated ion signal is converted into electrical. The signal is detected by an electron multiplier, and the detected signal is amplified and transmitted to the data processing system.

Shuanghuanglian oral liquid is composed of honeysuckle, skullcap and forsythia three traditional Chinese medicines. Chlorogenic acid, caffeic acid, baicalin and luteolin are the main active components of Shuanghuanglian oral liquid, and the structure is shown in Figure 3.8. In this experiment, HPLC-MS was used to identify the four active components of chlorogenic acid, caffeic acid, baicalin, and luteolin in traditional Chinese medicine Shuanghuanglian oral liquid.

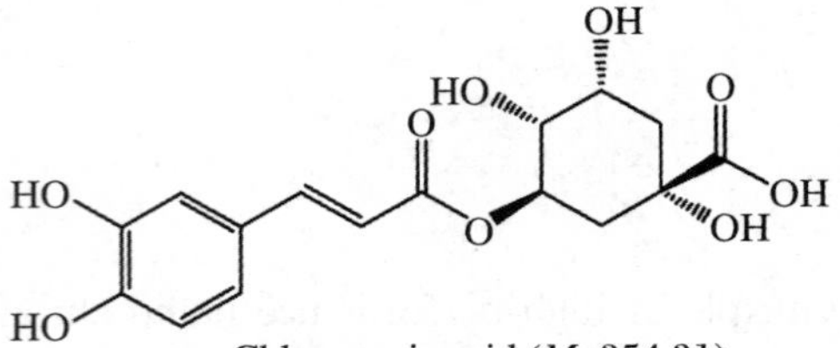

Chlorogenic acid (M=354.31)

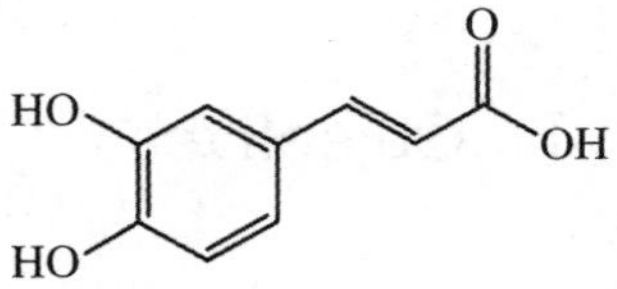

Caffeic acid (M=180.15)

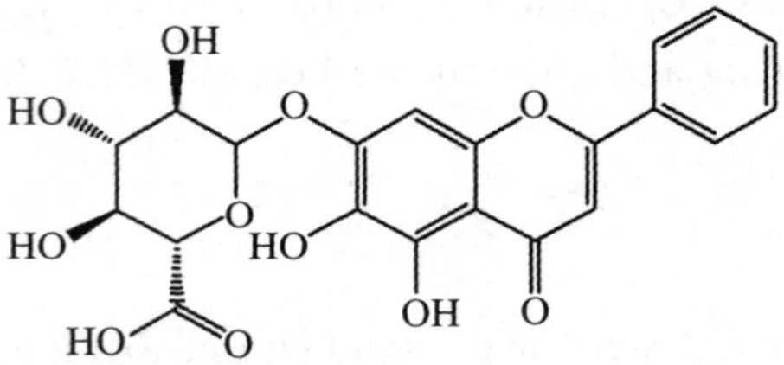

Baicalin (M=446.37)

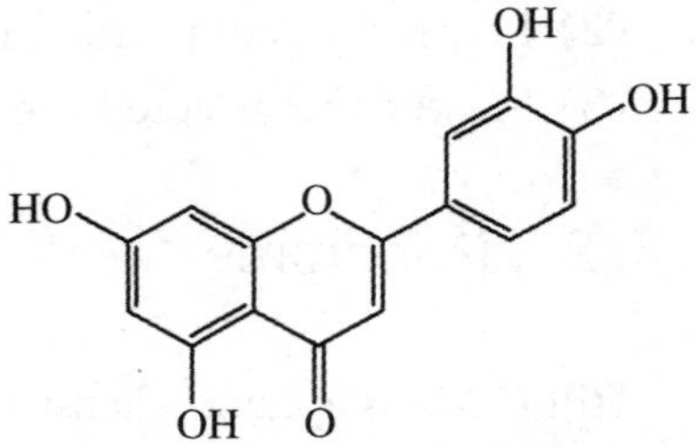

Luteolin (M=286.24)

Figure 3.8 Structures of chlorogenic acid, caffeic acid, baicalin, and luteolin

3.6.3 Instruments and Reagents

1. Instruments

High performance liquid chromatography-mass spectrometer (1100LC/MS Trap SL, Agilent).

2. Reagents

Methanol (chromatographic grade), formic acid (analytical grade), ultrapure water.

Sample: Shuanghuanglian oral liquid (commercially available).

Control substances: chlorogenic acid, caffeic acid, baicalin, and luteolin.

3.6.4 Procedures

1. Preparation of Solution

(1) Preparation of reference solution. Accurately weigh appropriate amounts of chlorogenic acid, caffeic acid, baicalin and luteolin and dissolve them in methanol to make a solution with a concentration of 10.0 μg/mL. 1 mL of each of four reference solutions must be mixed to prepare a mixed solution of the reference solutions.

(2) Preparation of sample solution. Dilute 100.0 μL Shuanghuanglian oral liquid with methanol to 10 mL, filter with 0.2 μm microporous membrane for HPLC-MS analysis.

2. Instrument Operating Conditions (Reference Value)

(1) Chromatographic conditions. Column: Johnson Spherigel C_{18} (250 mm × 4.6 mm i. d., particle size 5 μm); mobile phase: 0.3% formic acid in ammonium acetate solution (0.4mmol/L) (A)-methanol (B); gradient elution: 0～10 min (35% B), 11～25 min (65% B), 26～30 min (35% B); flow rate: 0.80 mL/min; column temperature: 25 ℃; injection volume: 20 μL.

(2) Mass spectrometry conditions. Split ratio 3 : 1, only about 0.2 mL/min enters mass spectrometer; ESI ion source, temperature 110 ℃; capillary voltage: 4.0 kV, cone voltage: 25 kV; the flow rate of nebulizer gas (N_2) and desolvation gas (sheath gas, N_2) is 50 L/h and 300 L/h respectively, sheath gas temperature: 300 ℃. ESI positive ion detection mode, time-based selected ion mode (SIM): 0～7 min, m/z 377.4; 7.0～12 min, m/z 181.0; 12～18 min, m/z 447.1; 18～25 min, m/z 287.1.

3. Sample Determination

(1) Measure the reference solution of chlorogenic acid, caffeic acid, baicalin and luteolin respectively, and obtain the chromatographic mass spectrum, in which m/z is selected as 377.4 ($[M+Na]^+$ ion peak of chlorogenic acid), 181.0 ($[M+H]^+$ ion peak of caffeic acid), 447.1 ($[M+H]^+$ ion peak of baicalin) and 287.1 ($[M+H]^+$ ion peak of luteolin) segment monitoring.

(2) Determinate the chlorogenic acid, caffeic acid, baicalin, and luteolin reference substance mixture.

(3) According to the peak area ratio of the sample and the reference substance in sample, quantitative analysis is used by the external standard comparison method.

3.6.5 Data Recording and Processing

(1) Assign the chromatographic and mass spectrometry peaks of chlorogenic acid, caffeic acid, baicalin and luteolin in the sample respectively, and calculate the separate effects.

(2) Calculate the contents of chlorogenic acid, caffeic acid, baicalin, and luteolin by external standard comparison method.

3.6.6 Cautions

(1) The volatile substances contained in the mobile phase are conducive for the conversion of liquid components into gas-phase ions in the ion source of mass spectrometry. Non-volatile salts (such as phosphate buffers or ion-pairing reagents) are not conducive for the conversion of liquid-phase components to gas-phase ions in the ion source of mass spectrometry. Therefore, the mobile phase of HPLC-MS should not contain non-volatile salts.

(2) When using the positive ion detection mode of HPLC-MS, in addition to the $[M+H]^+$ ion peaks of the components, there are often ion peaks in the form of $[M+Na]^+$, $[M+K]^+$, etc. The mass spectral signal intensity of various forms of ion peaks is greatly affected by the experimental conditions.

3.6.7 Questions

(1) Compared with MS, what are the advantages of HPLC-MS in pharmaceutical analysis applications?

(2) What are the main factors that affect the signal intensity of HPLC-MS mass spectrometry?

(3) Is it possible to use the negative ion mode of mass spectrometry to detect the analyte in this experiment? If so, what kind of ion peak appears?

第4章　设计性仪器分析实验

4.1　中药胆矾中 $CuSO_4$ 的含量测定

4.1.1　实验目的

（1）熟悉设计实验方案的思路及主要内容。

（2）总结所学分析方法的原理、步骤及应用。

（3）多方位考察学生对实验方法的掌握程度、对问题分析解决的能力以及书面表达能力。

4.1.2　实验要求

（1）教师提前三周将学生需要准备的实验内容告知学生，要求学生在十天内通过总结学过的分析方法及查阅文献资料，每人至少准备一份 $CuSO_4$ 含量分析方案。

（2）教师将学生准备的实验方案收集起来，进行认真审阅并告知学生实验方案中的不足之处，给学生提出改进的方向和思路。

（3）教师对所有学生的方案进行整体评判，归纳总结，根据实验室条件选择几个方案作为实验方案，告知实验员对实验进行前期准备。

（4）开放实验室，学生按照自己设计的实验方案进行实验。

（5）实验完毕后，整理实验数据并总结心得体会，写出实验报告。

（6）教师对实验报告进行评阅，帮助学生不断完善，提高学生自主分析解决问题的能力与实践水平。

4.1.3 可供选择的实验方案

1. 配位滴定法(返滴定法)

测定过程中首先以过量的已知浓度的EDTA标准溶液与Cu^{2+}相互作用,随后用已知浓度的Zn^{2+}标准溶液返滴定过量的EDTA,根据EDTA标准溶液的浓度和用量,得出Cu^{2+}浓度。此过程以二甲酚橙为指示剂。

2. 电位滴定法(配位滴定法)

该方法以电位滴定法确定滴定终点。用EDTA标准溶液滴定Cu^{2+},以Hg/Hg-EDTA为指示电极,饱和甘汞电极为参比电极。

3. 氧化还原滴定法(间接碘量法)

利用过量的KI与Cu^{2+}发生氧化还原作用生成一定量的I_2,再用$Na_2S_2O_3$标准溶液滴定反应生成的I_2,此过程以淀粉为指示剂。根据$Na_2S_2O_3$标准溶液的浓度及反应过程的用量,计算$CuSO_4$含量。

4. 重量分析法

将$CuSO_4$中的SO_4^{2-}用Ba^{2+}沉淀成$BaSO_4$,将沉淀经过滤、洗涤、灼烧至恒定,称量沉淀的质量,根据沉淀质量计算$CuSO_4$含量。

5. 紫外-可见分光光度法(标准曲线法)

制备一系列Cu^{2+}标准溶液,测定Cu^{2+}的吸光度,绘制Cu^{2+}的标准曲线得到线性回归方程。测定样品的吸光度A,根据线性回归方程计算Cu^{2+}浓度,进而得到$CuSO_4$含量。

4.1.4 注意事项

(1) 规范学生实验方案应包括的主要内容:① 实验的名称;② 实验目的;③ 实验原理;④ 仪器与试剂;⑤ 实验步骤;⑥ 数据与记录;⑦ 注意事项。

(2) 学生设计实验时应考虑某种方法测定待测组分时实验条件的选择及干扰的排除。

(3) 尽量将实验方案设计周全、条件简便可行。

Chapter 4 Designed Instrumental Analysis Experiments

4.1 Content Determination of $CuSO_4$ in the Traditional Chinese Medicine Bile Alum

4.1.1 Objectives

(1) Be familiar with the ideas and main contents of designing experimental programs.

(2) Summarize the principles, steps and applications of the analytical methods learned.

(3) Examine students' mastery of experimental methods, ability to analyze and solve problems, and written expression ability in multi-dimension.

4.1.2 Requirements

(1) The teacher should inform the students of the experimental content that the students need to prepare three weeks in advance and ask the students to prepare at least one analysis program about determination of $CuSO_4$ content by summarizing the analysis methods they have learned and consulting the literature within ten days.

(2) The teacher must then collect the experimental plans prepared by the students, review them carefully, inform the students of the deficiencies in the experimental programs, and give them directions and ideas for improvement.

(3) The teacher makes an overall evaluation of all the students' programs, summarizes them, selects several programs as the experimental program according to the laboratory conditions, and informs the experimenter to prepare for the experiment.

(4) In the open laboratory, students conduct experiments according to experimental programs designed by themselves.

(5) After the experiment, students should organize the experimental data, summarize the experience, and write the experimental report.

(6) Teachers ought to review the experimental reports to help students improve continuously and enhance their ability to analyze and solve problems independently and their level of practicality.

4.1.3 Schemes Available for Selection

1. Coordination Titration (Back Titration)

In the determination process, the excess EDTA standard solution of known concentration interacts with Cu^{2+}, and then the excess EDTA is back-titrated with the Zn^{2+} standard solution of known concentration. According to the concentration and dosage of the EDTA standard solution, the concentration of Cu^{2+} is obtained. Xylenol orange is used as indicator.

2. Potentiometric Titration (Coordination Titration)

This method determines the titration end point by potentiometric titration. Titrate Cu^{2+} with EDTA standard solution, take Hg/Hg-EDTA as the electrode indicator and saturated calomel electrode as reference electrode.

3. Redox Titration (Indirect Iodometric Method)

Excessive KI and Cu^{2+} are used for redox to generate a certain amount of I_2, and then the I_2 generated by the reaction is titrated with a standard solution of $Na_2S_2O_3$. In this process, starch is used as an indicator. According to the concentration of $Na_2S_2O_3$ standard solution and the amount of reaction process, the content of $CuSO_4$ is calculated.

4. Gravimetric Analysis

Precipitate SO_4^{2-} in $CuSO_4$ into $BaSO_4$ with Ba^{2+}, filter, wash and burn the precipitate to a constant state, weigh the mass of the precipitate, and calculate the $CuSO_4$ content according to the mass of the precipitate.

5. UV-Vis Spectrophotometry (Standard Curve Method)

A series of Cu^{2+} standard solutions are prepared, the absorbance of Cu^{2+} is measured, and the Cu^{2+} standard curve is drawn to obtain a linear regression equation. Measure the absorbance A of the sample, calculate the Cu^{2+} concentration according to the linear regression equation, and then the $CuSO_4$ content can be obtained.

4.1.4 Cautions

(1) Standardize the main contents of the students' experimental plan: a. the name of the experiment; b. the purpose of the experiment; c. the principle of the experiment; d. the instruments and reagents; e. the experimental steps; f. the data and records; g. cautions.

(2) When designing experiments, students should consider the selection of experimental conditions and the elimination of interference when determining the components to be tested by a certain method.

(3) Try to design the experimental scheme comprehensively, and ensure that the conditions are simple and feasible.

4.2 药物的有关物质检查实验

4.2.1 实验目的

(1) 掌握常见的仪器分析方法在有关物质检查中的应用。

(2) 掌握有关物质检查方法的建立和验证过程。

(3) 了解有关物质和药物合成过程中原料药和中间体。

4.2.2 基本原理

药物中的杂质按化学类别和特性一般分为:有机杂质、无机杂质及有机挥发性杂质(残留溶剂)。有机杂质主要包括合成中未反应完全的起始原料、中间体、副产物、降解产物等,可能是已知的或未知的。由于这类杂质的化学结构一般与活性成分类似或具有渊源关系,故通常又可称之为有关物质。有关物质研究是药品质量研究中关键性的项目之一。色谱法是有关物质检查的首选方法。用于有关物质检查的分析方法要求专属、灵敏。

4.2.3 可选药物

盐酸阿米洛利,吡罗昔康,黄体酮,樟脑。

4.2.4 实验方法

(1) 设计有关物质检查方法,写出实验方案和理论依据。

(2) 了解药物合成线路,根据实验方案,考虑实验室条件,选择合适的实验方法,写出操作步骤。

(3) 进行实际操作。对药物进行有关物质检查,制订合理的有关物质检查方法并进行方法学验证。

(4) 设计与《中国药典》不同的分析路线,并比较它们的优缺点。

4.2.5　实验指导

（1）设计实验前需查阅文献充分了解药物的合成途径，分析其合成所用的原料、中间体和副产物。

（2）可选择两种或两种以上的方法进行调研，最后确定最佳的检查方法。

（3）注意方法的专属性和灵敏度。

4.2 Related Substance Tests of Drugs

4.2.1 Objectives

(1) Master the application of common instrumental analysis methods in the related substance tests.

(2) Master the process of establishing and validating the methods for related substance tests.

(3) Understand the sources of related substances.

4.2.2 Principles

Impurities in drugs are generally divided into organic impurities, inorganic impurities, and organic volatile impurities (residual solvents) according to the chemical categories and characteristics. Organic impurities mainly include unreacted starting materials, intermediates, by-products, degradation products and so on, which may be identified or unidentified. Since the chemical structures of those impurities are generally similar to or related to the active pharmaceutical ingredient, they are usually called related substances. The research on related substances is one of the key projects in drug quality control. Chromatography is the preferred method for the test of related substances. The analytical methods used for the related substance tests should be specificity and sensitivity.

4.2.3 Optional Drugs

Amiloride hydrochloride, piroxicam, progesterone, or camphor.

4.2.4 Methods

(1) Design the examination method of related substance test and write the experimental protocol and theoretical basis.

(2) Learn about synthesis route of the drug, select the appropriate experimental

method according to experimental protocol and conditions of your laboratory, and write the operation procedure.

(3) Carry out the actual operation to examine the related substances of the drugs, develop a reasonable method, and perform methodological verification.

(4) Design analytical methods different from *Chinese Pharmacopoeia* and compare their advantages and disadvantages.

4.2.5 Experimental Guidance

(1) Before designing the experiment, it is necessary to review the literature to fully understand the synthetic pathway and stability of the drug and analyze the raw materials, the intermediates, by-products and degradation products.

(2) Two or more methods can be selected for investigation, and the best inspection method can be finally determined.

(3) Attention should be paid to the specificity and sensitivity of the method.

4.3 中药制剂的含量测定

4.3.1 实验目的

(1) 掌握中药复方制剂主要的含量测定方法。
(2) 掌握中药复方制剂含量测定方法的设计思路、条件优化和验证方法。
(3) 熟悉常用的仪器分析方法在中药含量测定中的应用。
(4) 了解中药复方制剂含量测定指标的选择原则。

4.3.2 基本原理

含量测定在中药制剂的研发、生产、工艺优化、质量控制和稳定性研究中起着重要作用。它能反映制剂中有效成分、毒性或指标成分含量的高低,衡量制剂工艺的稳定性和原料药质量的优劣,因此是最能反应药品内在质量的项目。

中药制剂多为以中医药理论和用药原则为指导而组成的复方制剂。因此在确定含量测定对象时,需要对组方有所了解。通常首选处方中的君药及贵重药建立含量测定方法。若上述药物的物质基础研究薄弱,或难以进行特征成分的含量测定,也可依次选臣药或其他药味的特征成分进行含量测定。对有大毒的药味要进行含量测定,若含量太低无法测定,应在检查项下规定限度检查项,或制定含量限度范围。测定成分应尽量与中医理论、用药的功能主治相近,并综合考虑生产工艺和功效。此外,处方中如含有化学药品也必须对其建立含量测定项目。

中药制剂常用的定量方法有比色法、分光光度法、薄层扫描法、气相色谱法、高效液相色谱法等。

4.3.3 可选药品

消渴丸、九一散、强力枇杷膏、小儿退热颗粒、连花清瘟胶囊、双黄连口服液、川贝枇杷糖浆、止喘灵注射液。

4.3.4 实验方法

(1) 根据选择的药物,查阅相关文献资料,写出简短的综述。
(2) 进行交流、讨论,选择含量测定指标成分。

(3) 写出实验方案、仪器与试剂、实验步骤。

(4) 完成所选药物的含量测定,包括样品的制备、仪器的操作、方法学验证、测定与计算,写出实验报告。

(5) 设计不同的分析方法,并比较优缺点。

4.3.5　实验指导

(1) 设计实验前需查阅文献资料,充分了解所选中药复方制剂的组成与制法。了解组方中君臣佐使和药效成分或有效成分。

(2) 需了解不同中药剂型含量测定常用的前处理方法,特别关注前处理对测定的影响。

(3) 可选择多种含量测定方法并进行比较,最后确定最佳方法。

(4) 选择合适的方法学验证内容,特别注意专属性和准确度指标。

4.3 Content Determination of Traditional Chinese Medicine Preparation

4.3.1 Objectives

(1) Master the common quantitative analytical methods of Chinese medicine compound preparations.

(2) Master the design ideas, condition optimization and validation of the quantitative analysis of Chinese medicine compound preparations.

(3) Be familiar with the application of commonly used instrumental analysis methods in the determination of traditional Chinese medicine.

(4) Understand the selection principles of content determination indicators for Chinese medicine compound preparations.

4.3.2 Principles

Assay plays an important role in the development, manufacture, process optimization, quality control and stability study of traditional Chinese medicine preparations. It can reflect the content of active components, toxic or indicator components in the preparation, and evaluate the stability of the preparation process and the quality of the crude drug materials, so it is the item that can best reflect the internal quality of the traditional Chinese medicine compound preparations.

Traditional Chinese medicine preparations are usually compound preparations guided by the theory of traditional Chinese medicine and the principles of drug use. Therefore, when choosing the object for assay, it is necessary to have some understanding of the principles for formulating prescriptions. It is usually preferred to establish the content determination method for the monarchs and precious drugs in the prescription. If there is a lack of research on the material basis of the monarchs and precious drugs or if it is difficult to conduct quantitation of the indicator components, the quantitative analysis can also be conducted on the characteristic components of the ministerial drugs or other drugs in the prescription. The content of highly poisonous medicine should be determined, and if the content is too low to be determined, the limit test should be specified, or the content limit range should be established. The determination of the ingredients should conform to the theory of traditional Chinese medicine and the function of medication as far as possible, and the production technology and efficacy should be comprehensively considered. In addition, if the prescription contains chemical drugs, it is also necessary to establish an assay for the chemicals.

The commonly used quantitativeanalytical methods of traditional Chinese medicine preparations are colorimetry, spectrophotometry, thin layer scanning method, gas chromatography, and high performance liquid chromatography and so on.

4.3.3 Optional Drugs

Xiaoke Pill, Jiuyi Powder, Qiangli Loquat Ointment, Antipyretic Granules for Children, Lianhua Qingwen Capsule, Shuanghuanglian Oral liquid, Chuanbei Loquat Syrup, Zhichuanling Injection.

4.3.4 Methods

(1) Read the relevant literatures and write a brief review about the selected drugs.

(2) Exchange and discuss, and then select the components for assay.

(3) Write the experimental protocol, instruments and reagents, and operation procedures.

(4) Complete the assay of the selected drugs, including sample preparation, instrument operation, methodology validation, determination, and calculation, and then write the experimental report.

(5) Design different analytical methods and compare the advantages and disadvantages.

4.3.5 Experimental Guidance

(1) Before designing the experiment, it is necessary to consult the literature to fully understand the composition and preparation of the selected

Chinese compound preparation. Learn about the monarch, ministerial, assistant, guide drugs, and therapeutic ingredients or active ingredients in the prescription.

(2) It is necessary to understand the commonly used pretreatment methods for the content determination of different Chinese medicine dosage forms and pay special attention to the influence of pretreatment on the results of assay.

(3) A variety of quantitative analytical methods can be selected and compared to screen out the best method.

(4) Select appropriate methodological validation items and pay special attention to the specificity and accuracy indicators.

附　　录

附录1　821型离子计的操作规程

(1) 将仪器选择开关拨至pX档,接上电极,电源开关置"开"位置。调节温度补偿至溶液的温度。

(2) 选择两种已知pX的标准溶液,例如溶液A为pX=5.00,溶液B为pX=3.00。选择的依据是被测对象的pX在两者之间。

(3) 将电极浸入较浓的一种标准溶液B中,调节定位旋钮使仪器显示为零。

(4) 将电极用蒸馏水冲洗干净,吸干外壁的水,插入较稀的溶液A中。如果电极的斜率符合理论值,则此时显示应为两种标准溶液的pX值的差(即ΔpX=5.00～3.00)。如果仪器显示值不符合ΔpX值,可调节斜率旋钮使显示器上指示值为ΔpX值。接着进行定位,用定位调节器使显示器指示出溶液A的pX值5.00,此时斜率补偿及定位完毕。在测量过程中该两旋钮应保持不动。

(5) 定位完毕后,用去离子水洗净电极,吸干外壁水,浸入被测溶液中,即显示出被测溶液的pX值。

附录 2　pHS-3C(A)型精密酸度计操作规程

(1) 连接 pH 复合电极,接通电源,打开开关,将功能旋钮置于 pH 档。

(2) 定位:将用蒸馏水清洗干净的 pH 电极插入 pH=7 的标准缓冲溶液中,调节温度补偿旋钮,使所示温度与溶液温度一致。再调节定位旋钮,使仪器显示 pH 值与该标准缓冲溶液在此温度下的 pH 值(查手册)相同。

(3) 调斜率:取出 pH 电极,用蒸馏水清洗干净并用滤纸吸干电极表面水分,再将电极插入 pH=4(或 pH=9)的标准缓冲溶液中。调节斜率旋钮,使仪器显示 pH 值与该标准缓冲溶液在此温度下的 pH 值(查手册)相同。

(4) 测量溶液 pH 值:取出电极,洗净,吸干。将电极插入被测溶液中,仪器显示 pH 值即为待测溶液的 pH 值。

(5) 测量完毕,将电极冲洗干净,放入电极保护液中。关闭电源。

注意事项:

(1) 无需精密测量时,在定位前将斜率旋钮拧到最大。

(2) 电极在放入另一溶液前需用去蒸馏水清洗,并用滤纸吸干电极表面水分。

附录 3　UV1000 紫外-可见分光光度计操作规程

(1) 连接电源,打开仪器电源开关,等待仪器自检通过。自检过程中禁止打开样品室。自检结束后,仪器进入主界面,即可进入正常测量。

(2) 光度测量:

① 在主界面上选中 A 项“光度测量”;

② 按“GO TO λ”键进入波长设置设定界面,用上、下键调整波长数值至测量波长,按“ENTER”键确认,仪器自动将波长移动至所需测定的波长值。

③ 打开样品室盖,将空白溶液和待测样品分别放置在比色皿架上,关上样品室盖。拉动拉杆,使空白溶液置于光路。按“ZERO”键对空白溶液进行自动校零。

④ 把待测试样拉入光路,可在当前工作波长下对样品进行测量,并记录相应的数据。

(3) 测量结束时,取出比色皿洗净,放好,关闭仪器,切断电源。

注意事项:

(1) 每台仪器所配置的比色皿不能与其他比色皿单个调换。

(2) 比色皿中放入液体的体积占其体积的 2/3～3/4,不能太少,也不能太满。

(3) 比色皿透光面用擦镜纸擦干净,以免造成测量的不准确性。

附录 4　CAAM-2001 原子吸收光谱仪操作规程

(1) 打开主机电源,打开仪器操作软件,点击“样品测量”栏下“元素选择”,选择待测元素后点击“下一页”进入“条件设定”。设置好参数(灯位、灯电流、负高压等)后仪器开始预热,预热时间至少为 30 min。

(2) 进行波长定位(波长偏离 0.3 nm 时用系统复位校正波长);再进行“波长扫描”(将起止波长设置为测量波长的 ± 1 nm,点击开始定位),定位结束后,点击“自动能量”,完成自动增益过程。

(3) 开启空压机(0.2～0.25 MPa),打开乙炔钢瓶阀门(0.06～0.1 MPa),并打开主机上的乙炔开关(注意空气及乙炔流量),点击“波长扫描”界面上的“点火”(应及时点火),完成仪器点火。

(4) 点火完成后,点击“下一页”,将标样个数改为绘制标准曲线所用的样品数目,测量次数为 3,并输入标准曲线浓度。

(5) 点击“空白吸收测量”,将毛细管放入纯水,如果基线离零点太远则点击“自动调零”,调整待基线稳定后点击“开始读数”,2 秒后点击“确认”。

(6) 点击“标样测量”,依次将毛细管放入配制好的标准溶液中,每次要等待基线稳定后再点击“开始读数”,所有标准溶液测试完成后点击“确认”,再点击“下一页”。

(7) 点击“浏览”,在“拟合方式”处选择“线性拟合”,点击“下一页”。

(8) 点击“空白测量”,将毛细管放入空白中,等待基线稳定后再点击“开始读数”,完成后点击“确认”。在“样品数目”栏输入待测溶液的数量。

(9) 点击“测量”,依次将毛细管放入待测溶液中,每次要等待基线稳定后再点击“开始读数”,所有待测溶液测试完成后点击“确认”。

(10) 测试完成后关闭乙炔和空气。

(11) 点击“结果输出”后点击“测定报告”,点击“文件”“打印”,即可打印实验结果。

(12) 关闭软件、仪器以及电脑。

附录 5　岛津 IR Affinity-1 型傅里叶红外光谱仪的操作规程

1. 开机前准备

开机前检查实验室电源、温度和湿度等环境条件，当电压稳定，室温在 15～25 ℃、湿度≤60%才能开机。

2. 开机

首先打开仪器的外置电源，稳定 30 min，使仪器能量达到最佳状态。开启电脑，打开仪器操作平台 IR Solution 软件，检查仪器的稳定性。

3. 制样

根据样品特性和状态，选定相应的制样方法并制样。固体粉末样品用 KBr 压片法制成透明的薄片；液体样品用液膜法、涂膜法或直接注入液体池内进行测定。

4. 扫描和输出红外光谱图

将制备好的 KBr 薄片轻轻置于样品架内，在软件设置好的模式和参数下测试红外光谱图。先扫描空光路背景信号(或不放样品时的 KBr 薄片，有 4 种扣除空气背景的方法可供选择)，再扫描样品信号，经傅里叶变换得到样品红外光谱图。根据需要，打印或者保存红外光谱图。

5. 关机

(1) 先关闭 IR Solution 软件，再关闭仪器电源，盖上仪器防尘罩。

(2) 在记录本上认真记录使用情况。

6. 清洗

清洗压片模具和玛瑙研钵。

附录6 岛津GC-2014气相色谱仪的操作规程

1. 开机前准备

(1) 检查载气系统泄漏、电源连接、载气种类、进样气垫等情况。

(2) 检查并核对色谱柱种类是否与标识卡标注种类相符合,色谱柱最高使用温度及色谱柱型号是否满足待分析样品的要求。

(3) 打开氮气、空气、氢气。

2. 开机

(1) 打开电源开关,待仪器系统自检通过后打开"Start GC"(或者从GC Solution工作站中打开)。

(2) 打开GC Solution工作站,单击通道1,进入实时分析窗口,从"文件"菜单中选择"打开方法文件",选择需要的方法文件并打开,然后点击"下载参数",将方法文件所设定的仪器参数传输到GC。

(3) 待"仪器监视器"小窗口中各项参数指标均处于"准备就绪"状态(绿色)且FID点火成功后,即可进行样品分析。打开"单次分析"进入分析操作界面,再打开"样品记录",输入样品名称和"数据文件"目录及名称,点击"确定"。再点击"开始"然后将样品注入进样口,按下GC面板上的"Start"按钮后即开始数据采集。

(4) 序列分析。对于配有自动进样器AOC- 20i的仪器可进行序列分析。打开"批处理",在编辑序列参数窗口中依次输入样品瓶号、方法文件、样品名称、样品保存路径等,待仪器准备就绪且基线走稳后,即可点击"开始"进行数据采集。

3. 方法编辑

(1) 从"文件"菜单中选择"新建方法文件",根据待分析样品的理化性质来设定合适的仪器参数(汽化温度、载气空气、氢气流量、尾吹流量、柱箱温度、检测器温度等)。

(2) 打开"文件"菜单中的"方法文件另存为",将新建好的方法文件以一个名称保存在合适路径的文件夹中。

4. 数据结果处理

(1) 打开工作站"再解析"进入数据分析:打开一个标样数据文件为参考,设置数据处理参数,另存这些参数为一个方法文件。

(2) 进入校正曲线:打开上述方法文件,对应校正级别增加标样数据文件,完成后保存。

(3) 进入数据分析:打开一个样品数据文件,加载上述校正曲线方法文件,保存计算结果到数据文件。

(4) 进入数据分析/报告文件:预览报告并打印。

5. 关机

(1) 仪器用毕关机时,先将柱温降至 50 ℃以下,进样口、检测器温度降至 100 ℃以下,可以专门建立一个"关机方法"来调用,降低仪器各单元温度。

(2) 完成降温后退出工作站,关闭电脑,关闭 GC 电源,关闭氢气、空气,最后再关闭载气。

6. 使用注意事项

(1) 及时更换进样口的进样隔垫,保证气路系统不泄漏。一般应在开机前更换,如开机后需更换则必须关掉载气后进行。

(2) 操作人员发现仪器工作异常时,应及时关机并报告,由专职人员进行检修。

(3) 氮气钢瓶压力降至 1.0 MPa,应予更换。

附录7　普析 LC-600 高效液相色谱仪操作步骤

1．开机前准备

(1) 在使用前应检查使用登记记录,确保仪器应处于正常可用状态。

(2) 检查供试品溶液及所需试剂溶液应准备齐全。

(3) 流动相应经过过滤、超声。

(4) 柱进出口位置应与流动相流向一致。

2．操作步骤

(1) 接通电源,打开电脑。

(2) 打开液相左下角的开关。液相右下角的绿色指示灯亮起时继续下一步(工作指示灯有三个:红灯(报警)、黄灯(等待)、绿灯(启动))。

(3) 打开桌面液相工作站,点击控制采集并启动,仪器进行自检。

(4) 单击右键设置检测器中的检测波长(如 $\lambda = 254$ nm),点击右键设置泵的流速(如 0.8～1.2 mL/min)。

(5) 打开排气阀(顺时针是开),将流动相里的气泡排出。

(6) 点击泵的开始开关,待将流动相里的气泡完全排出后,关闭排气阀。

(7) 点击检测器开关,计算机上图像自动走线,直到图像为直线时,开始用进样针进样(进样器后面有定量环,一次定量只能有 20 μL 样品进入,一般取样量为 40～60 μL)。

① 进样时将进样针进到底部,进样结束后立即将进样器旁的六通阀从 LOAD 打到 INJECT,等待图像。

② 当图像由直线—出峰—直线时,将六通阀从 INJECT 打到 LOAD。

③ 将图像保存至计算机,关闭检测器。若继续测量下一个品种,则重复上述步骤。

④ 采样结束后,先用正在使用的流动相冲洗半小时,再用纯甲醇或者纯乙腈冲洗 0.5 h。若流动相含有缓冲盐,则需先用 10%甲醇或者乙腈水溶液冲洗 1 h,再用纯甲醇或者乙腈冲洗至基线平直,冲洗完关闭检测器与泵的开关,关闭数据采集,关闭液相主机。

(8) 回到主页面,点击数据分析并启动,打开计算机里储存的需要分析的图像。

① 如图像有许多多余小峰,点击校正文件,选择积分界面,点击显示/隐藏事件表,再点击左上角事件类型,选择所要删除的积分所停留的时间,任意点击开始或结束即可删除,删除可重复多次。

② 点击校正参数，选择曲线方程（一般为线性曲线、包含原点，若图像出峰不好，可选择不过原点）。

③ 点击分析当前图谱，设定计算方法（一般选择归一化法或外标法）。

④ 使用外标法时需要制作标准曲线。

（ⅰ）点击新建校正任务；

（ⅱ）设置编号（编号数≥5 个）；

（ⅲ）根据标准溶液浓度设置化合物含量；

（ⅳ）计算曲线方程：$Y = aX + b$，根据标准曲线，通过已知的峰面积 Y，求浓度 X。

(9) 打印所需文件。

(10) 关闭桌面工作站，关闭计算机，关闭总电源。

(11) 用无水乙醇或甲醇清洗进样针，清理操作实验台。

(12) 填写高效液相色谱系统使用记录。

3. 注意事项

(1) 配制流动相时，有机相需要用有机膜过滤，水相需要用水膜过滤。

(2) 样品在进样前最好经 0.45 μm 或者孔径更小的滤头过滤。

(3) 使用完毕后，需将色谱柱和系统冲洗干净。

(4) 色谱柱若长时间不用，应冲洗干净后拆下两头封紧存放。